FARMERS' EXPERIMENTS

FARMERS' EXPERIMENTS
Creating Local Knowledge

James Sumberg
Christine Okali

LYNNE
RIENNER
PUBLISHERS

BOULDER
LONDON

Published in the United States of America in 1997 by
Lynne Rienner Publishers, Inc.
1800 30th Street, Boulder, Colorado 80301

and in the United Kingdom by
Lynne Rienner Publishers, Inc.
3 Henrietta Street, Covent Garden, London WC2E 8LU

Library of Congress Cataloging-in-Publication Data
Sumberg, J. E.
    Farmers' experiments : creating local knowledge / by James Sumberg
and Christine Okali.
        p.   cm.
    Includes bibliographical references (p.   ) and index.
    ISBN 1-55587-674-9 (hc : alk. paper)
        1. Agriculture—Research—Africa—On-farm.
    2. Farms, Small—Research—Africa—Citizen participation.
    3. Agricultural extension work—Africa.   I. Okali, C.   II. Title.
S540.053S86   1996
630'.724—dc20

96-28025
CIP

British Cataloguing in Publication Data
A Cataloguing in Publication record for this book
is available from the British Library.

Typeset by Letra Libre

Printed and bound in the United States of America

    The paper used in this publication meets the requirements
(∞) of the American National Standard for Permanence of
    Paper for Printed Library Materials Z39.48-1984.

    5   4   3   2   1

In memory of

Vera Edley
and
Robert Sumberg

Contents

List of Illustrations   ix
Acknowledgments   xi

1   Introduction and Overview   1

2   The Research Context   9

3   Farmers' Experiments: Concepts, Evidence, and Claims   31

4   Research Methods   55

5   Research Sites   65

6   Some Characteristics of Farmers' Experiments   87

7   Farmers, Experimentation, and Information   111

8   Farmers, Research, and Extension   147

Afterword   163

Appendix   167
References   171
Index   183
About the Book   186

Illustrations

TABLES

2.1   Estimated Distribution of Africa's Agricultural
     Population by Agro-ecological Zone   23
2.2   Suggestions for the Reorientation of Agricultural Research   25
3.1   Categories of Farmers   47
4.1   Classification of Examples of Farmers' Experimentation   59
4.2   Distribution of Interviews by Site   61
5.1   Characteristics of Research Sites   70
5.2   Agro-ecological Zones of Western Kenya   72
6.1   Examples of Farmers' Experiments from Africa   88
6.2   Characteristics of Farmers' Experiments   97
6.3   Examples of Farmers' Experiments from East Anglia   105
7.1   Distribution of Examples of Farmers' Experimentation   112
7.2   Experimentation and Socioeconomic Variables, All Sites   113
7.3   Experimentation and Gender   114
7.4   Experimentation and Age   115
7.5   Experimentation and Education   116
7.6   Factors Associated with Farmers' Experimentation   122
7.7   Categories of Named Individuals   131
7.8   Roles Played by Oft-Named Individuals   136
7.9   Characteristics of Named Individuals, Brong Ahafo   140
A.1   Characteristics of Respondents by Site   167
A.2   Characteristics of Respondents by Gender   169

FIGURES

2.1   Farmer Participatory Research in Relation to
     Other Key Themes in Development Discourse   26
2.2   Potential Synergy in Relation to the Characteristics
     of Farmers' Experiments and Formal Experimentation   30
6.1   Frequency Distribution of Topics, All Sites   89
6.2   Frequency Distribution of Topics, Individual Sites   90
6.3   Frequency Distribution of Topics in
     Ideas Competition, 1948–1994   104

6.4   Frequency of Spraying Entries in Ideas Competition   106
7.1   Frequency Distribution of Number of Times
         Individuals Were Cited as *Mhare/Changamuka*   134
7.2   Frequency Distribution of Particular Individuals Cited
         as *Mhare/Changamuka*   135
7.3   *Mhare/Changamuka* Farmers vs. Other Farmers   137

MAPS

5.1   Research Sites   69

Acknowledgments

Our research benefited from the interest and collaboration of many individuals and institutions. In Kenya, the link with the International Centre for Insect Physiology and Ecology was essential, and we gratefully acknowledge the assistance of Professor Fasil Kiros, Dr. Joseph Ssennyonga, and Mr. Andrew Ngugi. The Intermediate Technology Development Group (ITDG) in Zimbabwe provided a most congenial environment in which to work, and Mr. Kudaakwashe Murwira, Mr. Eddie Dengu, Ms. Monica Nyakuwa, and Mr. Colin Musvita deserve special mention. In Ghana, Dr. Dan Adomako, Dr. Ken Brew, Dr. K. Opoku-Ameyaw, and Mr. Francis Baek of the Cocoa Research Institute of Ghana and Mr. Kevin Gager, Dr. Kevin Waldie, Mr. Paul Amponsah, and Mr. F. W. Haizel-Cobbina of the Wenchi Farm Institute provided access to field sites and valuable feedback on our approach. We are particularly grateful to Dr. Kevin Waldie for his direct and insightful participation in the fieldwork in Brong Ahafo.

We also gratefully acknowledge the support of Ms. Felicity Proctor and Mr. Roger Smith of the Overseas Development Administration (ODA) and Ms. Cathy Watson of ITDG in helping to arrange the field visits. Dr. John Farrington and his colleagues at the Overseas Development Institute managed the grant under which the research was completed.

During the 1993–1994 academic year, students in the University of East Anglia's School of Development Studies' MSc program in Agriculture, Environment, and Development field-tested certain of our ideas about farmers' experimentation, and their patience and insights were most helpful. A number of colleagues commented on early versions of the original research report and manuscript, including Dr. Stephen Biggs, Dr. Arnoud Budelman, Mr. Alwyn Chilvers, Ms. Annika Dahlberg, Dr. Frank Ellis, Dr. David Gibbon, Dr. Elon Gilbert, Dr. Carla Roncoli, and Dr. Charles Staver. Their comments were essential and very much appreciated.

Finally, we are grateful to the many farmers who agreed to talk with us about their farming activities. Their goodwill was the key to the completion of the research.

This research was funded, in part, by the ODA of the United Kingdom. However, the ODA can accept no responsibility for any information provided or views expressed. That responsibility rests with us alone.

J.S.
C.O.

FARMERS' EXPERIMENTS

1

Introduction and Overview

Experimentation is probably as natural as conformity in traditional communities.

—Johnson 1972:156

The potential for reinvention should rather be encouraged; sponsoring agencies could offer local sites a variety of solutions or innovation components to perceived problems, and could then monitor the results of particular reinventions.

—Rice and Rogers 1980:511

The point is simply this: that local inventiveness is a resource that might be better integrated into the wider development process.

—Richards 1985:104

This book seeks to contribute to ongoing debates about different approaches to agricultural technology development and transfer. One central theme in these debates relates to the value and potential contribution of what is widely referred to as "indigenous technical knowledge," or local knowledge, and the processes by which this knowledge is created. Clearly, many individuals and processes are involved in the creation of local knowledge, and many sources of information feed into these processes. However, information about natural resources and, specifically, information about the technical and social dimensions of agricultural practice, regardless of its origin, must be tried or tested before it is incorporated into general practice. We will refer to this act of trying as "experimentation," and in our discussions with farmers we used the terms "to try" or "trying."

In this book, we set out to identify, characterize and put into context "experiments," or acts of trying, undertaken by farmers in Africa. In seeking to identify the characteristics of farmers' experiments, and the

1

factors associated with variation in these characteristics, we hope to help construct a more solid theoretical and empirical base from which alternative models for the interaction of formal research and farmers can be analyzed and evaluated.

At one level, the bounds of the debate about the interaction between farmers and researchers are clear and simple enough: Some people have suggested that farmers and agricultural researchers should communicate more; others have suggested that farmers and researchers should learn to interact "as colleagues"; and still others have proposed that there should be no "division of labor" between farmers and researchers. Each of these suggestions springs from a different understanding of the nature and processes of technical and social change, and, more specifically, the role, objectives, and methods of agricultural research. Each is promoted by individuals and groups with different interests and agendas, operating within different institutional frameworks. And finally, each of these suggestions has important implications—personal, institutional, and financial—for farmers, researchers, extension staff, development workers, academics, and policymakers.

This book specifically addresses the question of whether significant improvements in the performance of formal research and extension institutions can be expected if farmers' experiments are used as a specific and privileged locus around which a dynamic, "collegial" interface between formal agricultural research and farmers is created. In other words, it seeks an answer to the question of whether or not farmers' experiments represent an important and generally undervalued rural development resource. Thus, the book attempts to move beyond the general discussion of the now widely accepted need for greater farmer participation in formal agricultural research to an examination of the basis for one particular approach to participatory research. In examining this particular approach, we are concerned primarily with evaluating the claim that there is potential for synergy in bringing farmers' experiments much more centrally into formal agricultural research. In this case, we understand "synergy" to mean that the effect of combining farmers' experiments and formal experiments will be greater than the sum of the effects of each one alone (OED 1992).

We argue that the interest in greater farmer participation in formal agricultural research since the mid-1970s, and the associated interest in farmers' experiments, has had and continues to have important consequences for investment in, and models for, the organization and management of both agricultural research and rural development activities. In this way, work on farmers' experimentation can be seen to be directly linked with the wider policy discussion about the roles and limits of local participation in formal development activities and the implications this debate has for state institutions. Our concern is to help define the continuing role of the state in a context where the value of local contributions, in all their forms, are fully appreciated.

In moving through the book, and perhaps particularly in Chapter 3, which reviews a range of literature relating to farmers' experiments, the reader may well be tempted to question our judgment in choosing to focus on this topic. Are we not taking far too seriously what is essentially little more than rhetoric and posturing by a limited number of individuals and institutions with the objective of pushing a particular set of professional, institutional, or political agendas? It is certainly true that recent work on "farmer participatory research" shows clearly that there is an important disjunction between the rhetoric and the reality of much of what appears under this rubric (e.g., Okali, Sumberg, and Farrington 1994; Bentley 1994). In addition, the review in Chapter 3 demonstrates that much that has been said about farmers' experimentation is essentially ungrounded in either theory or empirical research.

However, it is critical to examine this mix of social and natural science, politics, ideology, policy, and development practice. We believe that the momentum for change in the relationship between farmers and agricultural researchers, which is partly the result of this mix, is having important impacts on the ability of research and extension institutions to contribute to the evolution and development of agricultural systems. The whole discussion of farmer participation in research, and the more particular suggestions as to the potential role of farmers' experiments, therefore, provides an excellent point from which to study the complex interplay among some of the academic, policy, practical, and ideological facets of current development discourse.

In examining these issues, we draw upon three main sources for information and insight. The first is the significant body of literature focusing on the relations between farmers and formal agricultural research and more recent writings on farmers' experimentation. The second source is a period of field research undertaken specifically to examine farmers' experiments in three countries in Africa. The third source that informs the book is our combined experience in agricultural research and development in Africa. This includes long-term work with national agricultural research systems (of Nigeria, Niger, Ghana, and The Gambia), international research centers (the International Livestock Centre for Africa [ILCA] and the International Institute for Tropical Agriculture [IITA]), and nongovernmental development organizations (CARE and Oxfam-America). Working within commodity, farming systems research, and institution-building programs has provided us with opportunities to consider issues that are directly relevant to those addressed in this book, including the relationship among farmers, researchers, and extension staff (Okali and Sumberg 1986; Sumberg and Okali 1988; Gilbert, Posner, and Sumberg 1990; Sumberg 1991; Okali, Sumberg, and Reddy 1994), and institutional and social relations as they affect both production and development interventions (Okali 1983; Okali and Sumberg 1985; Sumberg and Burke 1991; Painter, Sumberg, and Price 1994). In addition to these sources, we also draw upon a small body of work relating to farmers' experimentation in the United Kingdom (Rijal, Fitzgibbon, and Smith 1994; Lyon 1994; Knight 1995; Carr 1996). For although this book is primarily about smallholders in Africa, and the agricultural research and extension systems that are supposed to service them, the UK data add breadth to the analysis and raise important questions concerning the relationship between farmers and researchers in an entirely different context where commercial and large-scale production is the norm.

THE PARTICIPATORY RESEARCH STORY LINE

Farmers' experiments have long been of interest to a relatively small number of anthropologists and agricultural historians because of the logical and now widely accepted role of locally initiated experimentation in the complex of mechanisms by which farming practice and agricultural systems change. The experimental activities of farmers have also been a subject of study by those who are concerned with mechanisms by which new knowledge is generated and integrated into existing knowledge systems and practice. However, it was only with the more recent explosion of interest in participatory research that farmers' experiments began to enter the limelight. The specific interest in participatory research, as well as a host of associated themes within rural development discourse including participation, empowerment, local knowledge, civil society, and sustainability, has had an important influence on recent thinking about farmers' experiments. It is therefore fitting that our discussion of farmers' experiments is framed largely in terms of this more recent thinking.

There is now a well-established line of argument that is closely associated with the current interest in increasing the extent, and, more important, changing the nature, of participation by farmers in agricultural research. In its more extreme form, the argument can be outlined in ten logical and sequential steps, beginning with these assertions:

1. "Resource-poor farmers" in "complex, diverse, and risk-prone environments" did not benefit from, and indeed may have been harmed by, the technologies generally associated with the Green Revolution, and

they do not benefit from formal agricultural research and extension institutions and programs more generally.

2. Society and governments have obligations to address the needs of these individuals and regions; in addition, these individuals and their physical environments represent important and underutilized resources for national development.

However:

3. Normative models of formal agricultural research and extension will continue to be constrained in their ability to make significant contributions to resource-poor farmers in these areas, even after incorporating increased "client orientation" through modifications such as on-farm research and farming systems research; their apparent inability to tackle policy issues and inappropriate professional orientations severely limit the effectiveness of these models.

At the same time:

4. Farmers themselves have very detailed and valuable knowledge about their environments, which is expressed in their farming practices and reflects their needs, their own evaluative criteria, and so on.

5. Farmers actively do "experiments" and have their own "research traditions" through which they have, over centuries, introduced changes into their farming practice; there are some "research-minded" farmers who experiment as a major and deliberate activity.

But, unfortunately:

6. The positive contribution of farmers' own "research traditions" is being undermined by the progressive "commoditization of knowledge," the hegemony and the nonholistic orientation of formal agricultural research, and top-down, transfer-of-technology approaches.

7. Farmers' contributions are also being undermined by their weak experimental skills and poor access to new ideas, genetic material, and technologies, and by a lack of effective organization at the local level.

Therefore:

8. Research and development organizations must increase the level and change their mode of interaction with farmers; and farmers must be brought into the very center of the technology development process through the establishment of more "collegial" relationships.

9. These relationships will be facilitated by the transfer of better experimental skills to farmers and the creation of more effective local organizations.

10. This process will serve to "empower" resource-poor farmers by making them more confident and self-reliant and by giving them a clear role in the design of appropriate, locally adapted, and sustainable technologies.

It should be obvious that for many of the individuals and institutions actively promoting greater farmer participation in research, the story line outlined above is both too extreme and too simple. In fact, there is considerable disagreement in relation to certain of its elements, and these disagreements reflect, among other things, different professional and institutional imperatives. Nevertheless, this particular participatory research story line is important because of the power of its compelling simplicity, which, in certain institutional and political settings, can have important effects on decisions concerning the use of scarce research resources.

It is clear that there are a number of important assumptions underlying this story line. Although some of these have been highlighted, examined, and tested, many others have not. As will become evident in the following chapters, this is an important characteristic of much of the literature associated with both the general discussion of farmer participation in agricultural research and the more specific discussion of farmer participatory research and farmers' experimentation. One of the main objectives of this book is to examine some of these assumptions in more detail. In pursuing this goal, we maintain a somewhat narrow focus on farmers' experimentation and its integration into formal agricultural research for the more efficient and effective development and transfer of agricultural technology. Although we acknowledge the view on the part of some academics and many development workers that local empowerment and increased self-confidence and self-sufficiency (what we term "the social development agenda" of farmer participatory research) are central to the achievement of the objectives of increased farmer participation in agricultural research, we do not address these issues directly.

AN OVERVIEW

In Chapter 2, we sketch in more detail the context within which the discussion of the need for increased farmer participation in formal agricultural research developed. This context includes the institutions and paradigms associated with formal agricultural research and extension, as well as the more recent policy focus on poverty and agriculturally marginal areas. Efforts undertaken at the international, regional, national, and local levels to strengthen research and extension activities will be outlined, including moves toward greater farmer participation through farming systems research and "client-oriented" research more generally. Subsequently, we explore the practice of farmer participatory research and its relation to farmers' experimentation.

In Chapter 3, the existing literature on farmers' experimentation is reviewed in detail. This literature is drawn from a number of academic disciplines, as well as from the field experience of a variety of research and development organizations. The review identifies a surprising array of often quite contradictory characteristics and factors that have been associated with farmers' experimentation. We conclude that this literature is characterized by a lack of both conceptual clarity and empirical data, which has created a situation in which both policy and program implementation strategies are based as much on belief and ideology as on any understanding of underlying dynamics or processes. We argue that it is in this light that one can begin to understand the contradictions between the rhetoric and the reality of much ongoing farmer participatory research.

Our research methods are described in Chapter 4 and the research sites in Chapter 5. In Chapters 6 and 7, the results of the field research are presented: These are used first to analyze farmers' experimentation in terms of its subject matter, methods, sources of ideas, and outputs. Second, they are used to characterize the individuals who reported on these activities. A number of the key factors that appear to be associated with increased or decreased propensity to experiment are then identified. Chapter 7 ends with an analysis of information collected on networks through which agricultural information moves locally and of the link between these networks and farmers' experimentation.

Chapter 8 is a summary of the findings and a discussion of their implications. First, a number of hypotheses about farmers' experiments that were identified in Chapter 3 are reexamined in the light of our site-specific information. The central conclusion is that farmers' experimentation is widespread, an important part of everyday farming, and shares many characteristics with formal agronomic experimentation. Thus, through their experiments, farmers are involved in ongoing processes of local knowledge creation through site-specific learning, which, in the short term, results *primarily* in small adaptations to farming practice and, in the long term, contributes to the development of new farming systems. However, we conclude at the same time that although in many situations the arguments for greater participation of farmers in agricultural research are compelling and relevant, relatively little potential synergy will be realized through formal research and farmers' experimentation being more closely linked. In addition, because of the site-specific nature of the knowledge created through farmers' experiments, the claim that there is significant unrealized development potential associated with them, which could somehow be used to make an impact on a larger scale, is also called into question.

We then explore the implications of these findings for the larger debate concerning increased participation of farmers in formal agricultural research and various organizational and institutional models for agricultural research, extension, and development activities. What would be the likely benefits of a greater role for farmers' experiments, and what are the implications for the design and implementation of farmer participatory research and formal research more generally? The book concludes by returning to the broader discussion of how to increase the efficiency and effectiveness of processes of agricultural technology development and transfer, and the role that farmers might play within these processes.

2

The Research Context

The interest in greater farmer participation in agricultural research and the suggestion that there is significant potential for using farmers' experiments as a focus for that participation are rooted in a number of diverse debates and observations. Most of these have been reviewed in detail elsewhere (Bentley 1994; Okali, Sumberg, and Farrington 1994; Farrington and Martin 1988). In the sections that follow, attention is drawn instead to some of the key elements of these debates that serve as a backdrop for our work with farmers' experiments. We are specifically interested in the different views of the contribution and role of formal agricultural research and extension in agricultural change, approaches taken to making this research more effective, and the changing policy context of agricultural research. The last two sections of the chapter introduce the key elements of a critique of the practice of farmer participatory research.

FORMAL AGRICULTURAL RESEARCH
AND AGRICULTURAL CHANGE

Since the mid-1960s, significant bodies of scholarship and literature have developed around the topic of agricultural research in the developing world, and in recent years the situation in Africa has been highlighted. Major points of focus within this literature include assessments of the impact of investment in research (e.g., Echeverría 1990; Everson and Pray 1991); the political economy of research (Sims and Leonard 1990); links between research and other components and processes of the agricultural economy; models for the organization, management, and funding of research (e.g., Trigo 1987; Lynam and Blackie 1994), including, more recently, the particular challenges faced by "small" national research systems (Eyzaguirre 1992; Gilbert, Matlon, and Eyzaguirre 1994); and the relationship between government research structures and activities and non-governmental development organizations (Wellard and Copestake 1993; Farrington and Bebbington 1993).

If there is a general recognition that advances in technology have been and will continue to be critical for achieving increased agricultural productivity (with specific reference to Africa, see Delgado, Mellor, and Blackie [1987]), there is also overwhelming evidence pointing to positive rates of return to investment in formal agricultural research. Echeverría (1990) summarized the results of over 100 studies that covered a wide range of commodities in both developed and developing countries (although few were from Africa). Differences in methodology notwithstanding, almost all studies showed a positive return to investment in agricultural research, with calculated rates of return in many instances being over 50 percent. With specific reference to Asia, Everson and Pray (1991:358) came to essentially the same conclusion.

However, as one moves from these assessments based on changes in productivity to more people-centered analyses of the differential impact of research and technological change on society, any semblance of a consensus quickly disappears. Thus the Green Revolution experience in Asia has spawned a vast literature preoccupied with establishing the impacts on the rural poor, both positive and negative, of changes associated with the introduction of high-yielding rice and wheat varieties (see Lipton [1989] for a synthesis of these debates).

AGRICULTURAL RESEARCH IN AFRICA: HELP OR HINDRANCE?

There is a large and diverse body of formal institutions involved directly or indirectly in agricultural research, including universities, government ministries, private companies, and nongovernmental development organizations. These may have a commodity, product, or geographical (village, provincial, national, regional, or global) scope and funding base, and there are often important formal and informal links among them. These institutions are part of the "agricultural technology system," which Merrill-Sands and Kaimowitz (1990) define as comprising "all the individuals or groups working on the development, diffusion and use of new and existing technologies, the actions they engage in, and the relations between them." The system is described broadly as being made up of three "research sub-systems"—basic, applied, and adaptive—and a "technology transfer sub-system."

Much of the discussion is focused on two categories of institutions: the international research centers of the Consultative Group for International Agricultural Research (CGIAR) and national research centers. There is clearly much room for diversity within these two categories. The CGIAR is an association of over forty national governments, international organizations, and private foundations, which supports eighteen agricultural research centers around the world. It was founded in 1971 and maintains close links with major donors such as the World Bank (Anderson, Herdt, and Scobie 1988; Ravnborg 1992). Within sub-Saharan Africa, there are forty-six national research systems with total annual financing ranging from US$93 million for Nigeria to US$0.16 million for São Tomé and Príncipe (World Bank 1992:121). One of the central lessons emerging from the various studies of these national research systems is that experiences have been significantly different depending on the country, agro-ecological zone, or commodity involved (Merrill-Sands et al. 1992).

In effect, there are two parallel discussions about agricultural research in Africa. The first discussion is orchestrated largely by institutions such as the World Bank and the CGIAR (specifically through the International Service for National Agricultural Research [ISNAR]) and is concerned with assessing the impact of institutional forms and processes, especially systems of priority setting and overall management, on research efficiency and effectiveness (see Anderson 1994). The second is concerned primarily to make research more responsive to an activist, poverty-focused development agenda (Scoones and Thompson 1994a). These discussions converge in a "common knowledge" analysis vis-à-vis the contribution of, problems with, and remedies for formal agricultural research as a whole and national research systems in particular. This analysis is an important part of the context of our exploration of farmers' experiments because it is often invoked, in full or in part, to justify the push for greater participation of farmers in research. According to this common knowledge analysis, agricultural research in Africa (and in the developing world more generally) has been largely ineffective in meeting the needs of smallholders. The critique is developed in terms of (1) the researchers themselves, including their training, priorities, and methods; (2) the institutions and systems within which they work; and (3) the wider policy context within which research takes place.

Within this common knowledge analysis, researchers are portrayed as coming from the educated urban elite and are, consequently, assumed to have only limited firsthand knowledge of, and sympathy with, rural people (Sims and Leonard 1990:48). Since agricultural research is their job and their salary is not dependent on the impact of their work on agricultural productivity or rural livelihoods, they are simply motivated by the prospect of career advancement based, for example, on publications. It is argued, consequently, that many researchers would rather do "academic" experiments on a research station than face the discomfort and difficulty of working on the real problems of poor people in rural areas (Chambers 1989; Lipton 1985). Those individual researchers who may be interested in a different approach to their jobs are constrained by an institutional culture that forces either submission or departure.

At another level, the analysis suggests that many national research institutions, with their historical roots in the colonial period and an emphasis on plantation and export commodities, have only recently, and only halfheartedly, turned their attention to the wider range of food crops upon which the majority of rural people depend (Sims and Leonard 1990). At the same time, they continue to concentrate on technologies that demand inputs and resources that may be appropriate for a well-developed commercial sector but that are out of the reach of many farmers (Haverkort 1991; Chambers, Pacey, and Thrupp 1989:xvii). This orientation may continue into the future as the economic recovery and structural adjustment programs that are now in place throughout Africa put a high priority on increased production of agricultural commodities for export. Low adoption rates of technologies promoted by extension and the persistence of "top-down approaches" and "inappropriate" technologies are all seen as evidence of the continuing lack of effective influence or "demand pull" exercised upon research institutions by small-scale and resource-poor farmers.

We do not wish to suggest that many of the individual elements of this analysis are not important in particular situations, or that some do not have general significance. Our point is simply that, taken together, these elements present a powerful case for radical change within formal agricultural research. In fact, whether via this common knowledge analysis or via a more discriminating understanding, most writers agree that there is a pressing and widespread need to make agricultural research in Africa more effective, particularly in terms of meeting the needs of poor people in areas with resource constraints.

MAKING RESEARCH MORE EFFECTIVE

The challenge of making agricultural research more effective has been approached on a number of fronts over time. Perhaps one of the most significant and early efforts involved the strengthening and expansion of the overall research system through the addition of the network of regional "centers of excellence" under the CGIAR. The "international centers" within this system have brought new resources, both financial and human, into research and have had an important impact on the form and function of other components of the global agricultural research system, especially the national research systems. The research centers within the CGIAR view the national research systems as their immediate "clients." In addition to the direct transfer of technology and genetic material, they also provide training and facilitate networking activities (Ravnborg 1992).

The superior resource endowment of the CGIAR system has helped create and maintain a division of labor among the various components of the "agricultural technology system." For example, it is now widely suggested, particularly from within the CGIAR system and its funders, that the international centers should be concerned with basic and strategic research, while national systems, particularly those operating under severe resource constraints, should focus primarily on applied and adaptive research. This division is reinforced through the training and networking activities of the CGIAR centers, which serve, among other things, to introduce the centers' technologies into local research programs and extension activities. Indeed, some of the centers' networking activities give new meaning to the notion that national systems are the clients of the international centers. In this context, the term "client" is usually used to indicate that the national systems receive the products or services of the CGIAR centers. However, networking activities that provide small amounts of funding to national researchers specifically to test a center's technologies, often using a protocol approved by the center, effectively result in the national researcher becoming the "client" of (i.e., under the protection of, or dependent upon) an international "patron."

Another avenue pursued in attempting to make agricultural research more effective has been to strengthen the national systems through human resource development and restructuring exercises. Thus, utilizing both recurrent budgets and special projects, there has been a continuing effort to develop human resources through degree-level, postgraduate, and specialized training. At the same time, the push to restructure has focused on improving cost effectiveness and financial accountability. ISNAR, itself one of the CGIAR centers, has played a major role in this institutional reorganization, while the World Bank has provided much of the funding (Busch and Bingen 1994). It must also be remembered that these changes in national agricultural research systems mirror much broader trends in thinking and policy concerning the size and role of government in the provision of rural services.

In our view, perhaps one of the most important developments since the 1970s has been the various initiatives to increase the "client orientation" of national research systems and programs (Biggs 1989; Merrill-Sands et al. 1991). The most significant of these initiatives in terms of funding and staffing has been the promotion of farming systems research. Farming systems research, which has its roots in farm management economics and systems theory, is based on the assumption that the development and testing of agricultural technology must take place in the context of specific farming systems; it acknowledges the fact that processes of, and decisions affecting, technology development and use are necessarily socially embedded. Farming systems research programs, therefore, have sought to legitimize and institutionalize client involvement in the research process, including problem identification and technology evaluation, with some programs working toward these ends with farmer groups (Norman et al. 1988; Heinrich 1993). Associated with these efforts are a number of other strategies that seek to integrate farmers into research management functions. It was in this light that some programs and research systems added farmer representatives to research advisory committees (Biggs 1989).

The development of farming systems research programs and the way these programs, and the principles of farming systems research more generally, have been incorporated into existing research systems have been highly variable. Although it is likely to be with us for a long time, and a systems approach is now widely advocated for understanding smallholder production activities, farming systems research has been criticized for its failure to achieve expected outputs (i.e., increased welfare of marginal groups including the poor and women) and to fully integrate a political perspective (Scoones and Thompson 1994a; Biggs 1994). More recent attempts to increase the degree of client orientation have, therefore, taken place largely outside the framework of farming systems research and include a very broad range of approaches and activities that all claim the title "farmer participatory research."

This shift also reflects, in part, the increasing role of nongovernmental organizations (NGOs), with their broader development agendas, in the design and delivery of rural development services, including agricultural research and extension. From a policy perspective, the growing interest in farmer participatory research reflects the way in which all policy arenas, such as natural resource management or health and population, are now expected to address social development concerns of participation and empowerment, in addition to more conventional agendas of agricultural technology development and transfer or disease and population control.

It is absolutely critical to realize that there is little that is new, much less radical, in the proposition that farmers have a role to play within agricultural research. In fact, farmer participation is a well-established principle within at least some agricultural research traditions. For example, in what may be one of the earliest examples of client-driven, participatory diagnosis and priority setting within a formal research context, Box (1967:44) describes the beginnings of the U.S. government's range research effort in southwest Texas. In 1886, following the degradation of range vegetation and a series of hard winters during which cattle mortality was high, the Department of Agriculture's Division of Agrostology sent letters to 2,000 cattle ranchers requesting their views about range conditions. An assistant agrostologist was subsequently sent to Texas to gather firsthand information from the ranchers. The following year, the government embarked on a program of range research by establishing two research stations. If they were to take place today, these activities would likely be referred to as "participatory problem diagnosis," "rapid rural appraisal," and "on-farm, on-station linkages."

There are many other examples of the participation of farmers in various aspects of research, from holding seats on funding and oversight committees to visiting demonstration plots and attending experiment station field days. In East Anglia in 1908, farmers founded what is today known as the Morley Research Centre. Since that time, farmer members have played central roles in both the funding and the direction of its research activities (Hutchinson and Owers 1980:12; McClean 1991). Indeed, it would be difficult to explain the dramatic impact of agricultural research in much of the world if one assumed only limited or one-way interaction between research and farmers. At the same time, it cannot be seriously suggested that all *regions* or all *strata* of the farm population have participated in, or benefited equally from, agricultural research.

If the principles of client orientation and farmer participation are so well established, one might ask, what can account for the large gap that is so often assumed to exist between researchers and farmers in some parts of the developing world? Is there something special about the historical context, or about present-day institutions, researchers, or farmers, that prevents beneficial contact and exchange? In response to these questions, many observers return to the litany of factors at the heart of the common knowledge analysis outlined above: colonial heritage, historical emphasis on export crops, extreme social distance between researchers and farmers, resource limitations, and diversity of agroclimates. Clearly, assumptions associated with worldviews that juxtapose modern and primitive, civilized and savage, and educated and ignorant did not, and indeed do not today, create fertile ground for the growth of an open, interactive relationship. On a more practical note, fundamental misunderstandings of the social, cultural, economic, and psychological meaning of "farmer" and "farming" in the context of diversified rural livelihoods in Africa have also contributed to the problem.

However, the argument that formal research has yielded little of value can easily be taken too far. It would be wrong to assume that farmers in Africa, even resource-poor farmers, have not benefited from agricultural research. New crop varieties are widespread, as is the use of a range of agronomic practices and tools that originated with, or were modified by, formal research. In some cases, it may be argued that small-scale, resource-poor producers have simply been clever in making use of

technology developed for other users, and this is essentially the suggestion of Eicher and Rukuni (1994) in relation to the successful use of
short-cycle maize hybrids by communal farmers in Zimbabwe. Otherwise one has to conclude either that formal research was blessed with
blind luck, or that it actually did succeed in identifying some of the
needs, interests, and capabilities of some part of the population of resource-poor producers.

In fact, the discussion of the effectiveness and impact of agricultural
research, which was touched upon earlier, is fraught with difficulty. Recent work on the impact of over thirty years of investment in maize research in Africa highlights the fact that farmers use the products of agricultural research in unintended ways (Gilbert et al. 1994; Gilbert 1995).
For example, although many farmers have adopted the new maize varieties, for some, their total maize production has remained static as they
use the greater productivity of the new varieties to shift land and labor
to other crops and economic activities. Gilbert et al. refer to this as the
"impacts iceberg" and conclude that "a significant proportion of impacts
are associated with improvements in returns to labour, reduction in neg-
atives, and reallocation of resources that are not readily visible through
available statistics" (p. 95). In the context of complex and diversified
livelihood strategies, the observation, quantification, and attribution of
the impacts of agricultural research are highly problematic. Unfortunately, in the face of these difficulties, the common knowledge understanding tends to assume no positive impact at all.

EXTENSION, COMMUNICATION,
AND TECHNOLOGY TRANSFER

Improvements in the effectiveness and impact of research have also
been sought via agricultural extension (part of the so-called technology
transfer subsystem). Whereas extension is primarily viewed as being responsible for the transfer of information from formal research to client
groups, it is also expected to provide research with insights into farmers'
needs and their views on research outputs. Thus, functional research-extension linkages are a priority for effective agricultural technology systems, and alternative strategies and mechanisms for linking research and
extension have been the subject of considerable study and discussion
(Ewell 1990; Kaimowitz, Snyder, and Engel 1990; Eponou 1993).

At the time of independence, national agricultural extension systems in Africa, like the agricultural research services, were focused primarily on export commodities. Multifunctional extension systems covering food crops and distributing inputs including credit emerged in the

1960s and 1970s alongside the expansion of rural development projects. By the late 1960s, however, the World Bank was already expressing concern over the lack of attention being given to extension's technology transfer role. Its concept of a more specialized, "freestanding" extension service was first tested in Turkey in 1967 in the form of the Training and Visit (T&V) system. By the end of the 1980s, Africa had become a major focus of T&V extension projects; and by the 1990s, these projects were being modified in a variety of ways to incorporate more participation by farmers.

Thus, since the early 1970s, significant investments in national agricultural extension systems in Africa have been made (World Bank 1994), hand in hand with the various attempts to make researchers more cognizant of, and responsive to, the needs of resource-poor farmers. The World Bank has funded small-scale programs since that time, and presently, thirty countries have World Bank–funded variants of T&V programs (Howell 1988; Hulme 1991; Bagchee 1994).

As operational models of extension have evolved, so has what Röling (1988) refers to as "extension science." It has been dominated by a sequence of different themes, from effective dissemination of information and the strategic use of different formal channels for changing behavior in the 1960s, to more recent interest in alternative sources of information, modes of learning, and knowledge creation. Recent discussion has highlighted processes of information exchange, especially between what are described as "informal and local" and "formal and central" knowledge systems, and the effect of institutional variables on these processes. If the importance attributed to each of these themes has changed over time, they all remain identifiable within the current literature (see Wallace, Mantzou, and Taylor 1996).

As is the case with research, there are many different actors who are either potentially or actually involved in these processes of information exchange. Thus information comes from "local" sources (e.g., family, friends, and traders) and through "informal" channels (e.g., gossip and simple observation), as well as from "central" sources (including government extension agents and researchers) and through "formal" channels (including extension bulletins, research station visits, and demonstration plots). Many different institutions are involved in information exchange, including state extension, nongovernmental development organizations, commodity programs run by parastatal organizations, and farmers' cooperatives. Within the context of such a framework comprising multiple sources, channels, and institutions (Biggs 1990), all actors are potential givers and receivers of information, and behavior change becomes a concern to be addressed at different levels by all the individuals, groups, and institutions involved.

Despite the fact that extension is now frequently analyzed within such a broad framework, much of the discussion continues to be concentrated on the workings of the formal (usually state) extension systems. Within this discussion, similar to that pertaining to the research system, considerable attention is given to an examination of the behavior of the institutions involved vis-à-vis one category of end clients, the producers, and the assumptions and institutional imperatives underpinning that behavior.

Clearly, the expectations of the outputs of formal extension are high, and much of the literature assesses the ability of state systems to meet these expectations (World Bank 1994). In spite of acknowledged data problems, evaluations have frequently been negative and point to the inability of national systems to satisfy either the producers or their other principal client, research. National extension services have been described as "ponderous, expensive, alien," often with "inequitable channels and organisation for disseminating agricultural information" (McCorkle 1989), and characterized by "rigid institutional frameworks and unwillingness to listen" (Engel 1990). The main criticism of the state systems revolves around the models or premises on which their behavior is assumed to be based.

The first model, and what many observers assume to be the current model, is linear, top-down, and hierarchical, concerned with the transfer of information from research to farmers (Rogers 1993), with no explicit acknowledgment of the value of local knowledge. This is essentially the same image given by Biggs (1990) in his description of the "central source" model of innovation. Engel (1990) views the consistent underrating of "non-scientific research-based knowledge" as one of the fundamental reasons why results from extension have generally been disappointing. He therefore argues that the role of extension is to facilitate, in a specific situation, the 'fusion' of farmer knowledge, on the one hand, and research-based, technical knowledge, on the other. Others see extension as providing a bridge between new knowledge and traditional knowledge (Saito and Spurling 1992). In either case, the implication is a need for a more explicit appreciation of the richness and diversity within what is referred to by some as the larger "agricultural knowledge and information system" (Röling 1988).

The general failure of state systems to acknowledge the existence and importance of multiple sources of information and to incorporate these into their programs of work presumably reflects, in part, the fact that government extension services were set up with the explicit task of disseminating the results of government research. National research and extension organizations also tend to centralize and standardize information in order to provide simple, all-inclusive solutions or technical recommendations. As we have discussed previously, the fact that a number of possible solutions or options for different situations might have been identified, or that research might not be in a position to provide all the options at once, is generally overlooked in the process of formulating recommendations (Okali, Sumberg, and Reddy 1994). This practice of seeking to make recommendations that have wide applicability can be related to the need for state systems to address a national audience. With these points in mind, it must be acknowledged that the stereotypical patterns of organizational behavior so often associated with extension do not necessarily reflect a simple lack of understanding of the value of farmer knowledge and practices.

However, with the increasing recognition of the value of local knowledge, the task of extension is no longer viewed as simply the transfer of messages relating to new technology. Rather, state extension systems should now provide specific new information only after assessing local needs and local knowledge. This approach is referred to by Mc-Corkle, Brandstetter, and McClure (1988) as a "second stage model" and by Rogers (1993) as "second generation" extension. With these models, the problem-solving function remains in the hands of the research-extension hierarchy, and the process continues to be essentially linear and concerned with the dissemination of information from research. However, these models are associated with greater expectations on the part of some observers, with new information enabling local people to take more control over their own lives (e.g., Riches et al. 1993). In this way, they are assumed to have the potential for "strengthening customary patterns and networks for learning"(Rogers 1993).

EXTENSION AND PARTICIPATION

The general response from national extension systems to mounting criticism has been to incorporate more participatory methods. The second-generation models referred to above can be seen in this light. Yet the response has not, in general, involved the creation of "advisory partnerships" that involve farmers in setting goals or been directed toward action programs aimed at strengthening farmer-to-farmer communication (McCorkle, Brandstetter, and McClure 1988).

One of the more complex examples of a participative, feedback model arose from the farming systems research framework: The "farmer-back-to-farmer" model begins with the explicit assumption that the clients need to define research priorities (Rhoades and Booth 1982). However, extension was not initially incorporated into the farming systems framework (Sowers and Kabo 1987), even though it was acknowledged that, without it, it was not possible for researchers to work closely with farmers on anything but the most limited scale. The T&V model, on the other hand, was designed as an extension system with explicit emphasis on research extension linkages, which, in addition to improving extension, are expected to contribute to the goal of a more informed and responsive research system. Within T&V, however, field agents continue to be viewed as teachers, and the whole program of activities is based on the assumption that there is a finished product ready to be passed to farmers. T&V was designed to deliver selected and timely messages to farmers with strict regularity, rather than to create a dialogue with farmers.

Even more than in the case of research activities, extension activities have been taken over by NGOs as participative approaches have become more popular and state systems, suffering from chronic underfunding, appeared to hesitate. NGOs are generally expected to be more successful in implementing participatory extension systems because of their closer link with the grassroots (Farrington and Bebbington 1993). The most common participatory extension technique used is the strengthening of farmer-to-farmer information transfer, and some NGOs have provided training to state agencies in the use of such techniques. Nevertheless, although the farmer-to-farmer model is described as participative, it continues to be concerned primarily with information transfer. It is not, therefore, based on a principle of fusion or knowledge integration as described by Engel (1990).

The expanding role of NGOs in rural areas has therefore created alternative, complementary, and, in some cases, competitive extension activities. These programs and activities also place demands on, and can provide feedback to, formal research institutions and can thus be additional vehicles for the participation of farmers in research. It has been suggested that links with development organizations can be particularly important for small or severely resource-constrained research systems (Gilbert, Matlon, Eyzaguirre 1994). In recognition of the strength of NGOs, there is a movement toward partnerships. Chowdhury and Gilbert (1996), for example, describe agreements where government and nongovernment staff members work together based on an agreed division of labor, with government staff providing technical skills and nongovernment staff providing links with the rural community through training and motivational activities.

It is argued in much of the participatory literature that state extension and research services are limited by their use of "scientific" methods, their hierarchical structures, and their professional cultures. It is clear that in many cases they are also limited by the way in which public institutions operate in rural areas. For example, government extension staff are routinely relocated and are renowned for their heavy responsibilities in terms of geographical coverage, while more often than not being seriously constrained by a lack of transport and other logistical support. Special programs funded through the World Bank have tried to address this problem, but as projects and funding come to an end, the problem reemerges. NGOs, in comparison, are often involved in local programs over a relatively long period of time, which potentially enables staff to build strong alliances with local populations, relationships that public institutions generally seek to avoid. These longer-term relationships fit more with the NGOs' overall objective of social development and "radical social action through local-level interventions" (Brown 1994).

In a recent review of the extension system in Bangladesh, Chowdhury and Gilbert (1996) suggest that some of the criticism of extension systems reflects simply unrealistic assumptions regarding the willingness and ability of different organizations to make changes and to work together. We suggest that this observation is equally relevant to the discussion of agricultural extension in Africa.

CHANGING FOCUS:
POOR FARMERS AND MARGINAL AREAS

One important change in the context within which much agricultural research in Africa takes place is the now widely accepted emphasis of rural development policy and programs on food security and poverty alleviation (Lipton and Maxwell 1992). The so-called new poverty agenda has important implications for both agricultural research institutions and development organizations. Lipton and Maxwell (1992), for example, suggest that compared with the 1970s, poor people in the 1990s are more likely to be living in resource-poor areas and to be refugees or displaced. They also indicate that as a consequence, research will need to move away from its traditional emphasis on high-yielding varieties and be more cognizant of "poor people's problems," such as poor plant establishment and losses from weeds, birds, and rats.

Chambers, Pacey, and Thrupp (1989:xviii; Chambers 1989:182) were instrumental in popularizing the phrase "complex, diverse, and risk-prone" to describe the environments in which the rural poor live. These areas are normally considered to have relatively low agricultural potential and are often contrasted with high-potential areas, which were the focus of the Green Revolution. These "marginal areas" have some combination of low and unreliable rainfall, poor soils, and hilly topography, and it is claimed that they have long been neglected by national research and extension systems. According to one view, this neglect is a reflection of both their physical isolation, which makes them inaccessible to centrally administered research systems, and the inappropriateness of the aims and methods of conventional research (Chambers and Ghildyal 1985:18; Farrington and Mathema 1991). Going even further, Haverkort (1988) suggests that formal research is itself at least partly responsible for the magnitude of the challenge posed by these low-potential areas because the definition of particular agro-ecological situations as having "low agricultural potential" is a result of both the "male Western bias" of formal research and past research investment decisions.

Apart from issues of equity, the call for increased research expenditure in marginal areas is based on the proposition that if these areas are reassessed according to different criteria, such as the proportion of useful biomass generated from common property resources, a different view of relative potential and marginality will emerge (Farrington and Mathema 1991). There is also an assumption that substantial and sustainable increases in productivity will result from using more participatory research methods and from institutional reforms (Chambers, Pacey, and Thrupp 1989; Farrington and Martin 1988; Farrington and Mathema 1991).

However, the relationships between poverty and agro-ecological potential remain obscure as there have been few attempts to actually map the relative or absolute occurrence of rural poverty in Africa in relation to agroclimates. The discussion is also confused by broad generalizations: Chambers (1989:xvi), for example, suggests that agriculture in "most of sub-Saharan Africa" can be considered complex, diverse, and risk-prone. Jahnke (1982:235) provides what he calls "rough estimates only" of the distribution of Africa's agricultural population by ecological zone: 10 percent and 28 percent are in the arid and semi-arid zones, respectively (Table 2.1), although the distribution of poverty is not mentioned. Mellor (1988) argues that rural poverty in Africa is distributed evenly between areas of "low" and "high" agricultural potential.

The problem is that some observers argue that areas defined as having low potential will be the most difficult in which to achieve rapid increases in productivity, and it is logical that research should therefore concentrate on higher-potential areas (Mellor, Delgado, and Blackie 1987:362). On the other hand, there are certainly those who suggest that there are major opportunities for increasing agricultural productivity in marginal areas (Byerlee 1994:225; Lipton 1994:611) and that agricultural research in these areas will be crucial for the success of the new poverty agenda (Lipton and Maxwell 1992:15). Nevertheless, Byerlee (1994) indicates that in marginal areas new crop varieties and management practices will have to be introduced *simultaneously* with improvements in input supply and "strong adaptive crop management research and extension systems" (p. 225). It is clear that in order to realize op-

Table 2.1.   Estimated Distribution of Africa's Agricultural Population by Agro-ecological Zone

Ecological Zone	Annual Rainfall (mm)	Agricultural Population (×1,000)	Agricultural Population (%)
Arid	< 500	24,853	10
Semi-arid	500–1,000	65,735	28
Subhumid	1,000–1,500	59,442	25
Humid	> 1,500	50,307	21
Highlands		37,971	16
Total		238,308	100

Source: Jahnke (1982:235), 1979 data.

portunities for increasing agricultural productivity in many low-potential areas, a number of institutional and political changes that are outside the realm of agricultural research will be needed.

The net result of a policy focus on poverty alleviation in marginal areas is creating a serious challenge for formal research in Africa. Institutions that are almost universally acknowledged to be weak, understaffed, and underfunded are being asked to develop productive (and now sustainable) technologies for situations that can only be considered difficult. To add to the challenge, the potential users of the technologies have few disposable resources and generally poor access to supporting services.

In terms of extension, it is fairly widely agreed that in contexts where farmers are commercialized, powerful, organized, and concentrated, information flows from research via extension almost automatically. Röling (1988) argues that it is only in situations such as these that extension can fulfill its role. In contrast, Ford and Babb (1989) suggest that there is a wider range of situations within which state systems can function effectively, and these situations should therefore be privileged targets. They specifically cite crises programs (e.g., following drought), "new technology" areas (including low-input sustainable agriculture), and anything requiring multidisciplinary inputs and complex decisions.

It is, of course, in the agriculturally less productive areas that NGOs in Africa are most active. Often initiating work along emergency or relief lines, their programs eventually shift to development issues. Emphasizing grassroots participation and community empowerment, many NGOs have deep-seated animosity toward government structures, including formal research and extension institutions. By the same token, they tend to have a natural sympathy with much of the current discourse that justifies and promotes more "participatory approaches," including farmer participatory research.

This skepticism toward formal research is fueled by populist discourse and the current emphasis on indigenous knowledge and knowledge systems. When these are brought together with the NGOs' more fundamental interests in participation and empowerment, the conclusion is that nothing less than a radical reorientation of agricultural research is required (see Table 2.2). In addition, there is an implicit assumption that valid contributions to agricultural research can be made from many quarters. The prerequisite to do research is not one's understanding of a particular discipline and its associated theory, methods, and so forth, or, indeed, the depth of one's local knowledge but, rather, one's goals and beliefs. The paramount position of both participation and indigenous knowledge, and the accessibility of the tools and techniques associated with participatory rural appraisal, make, in the eyes of some development workers, formal research redundant (if not a very real hindrance).

Thus, we suggest that agricultural research is caught on the horns of a dilemma: On the one side, it is burdened with "responsibility" for supporting the development of more productive agriculture in the most difficult rural areas; on the other, it is critiqued for its traditions and methods and pushed to be less formal and directive, more participative, and so on (see, e.g., Chambers [1989] and his more recent calls for a "new professionalism" [1993]). All this is taking place in the context of restructuring, downsizing, and the search for greater cost-effectiveness.

FARMER PARTICIPATORY RESEARCH

The level of interest in greater participation of farmers within formal research is evident in the growing number of books and workshops that have attempted to collate, review, and analyze recent work in this field (Farrington and Martin 1988; Chambers, Pacey, and Thrupp 1989; Amanor 1990; Haverkort, van der Kamp, and Waters-Bayer 1991; de Boef, Amanor, and Wellard 1993; Scoones and Thompson 1994b). In our own review of this material, we noted the wide range of approaches and activities that are associated with the label "farmer participatory research" or any of a number of other labels including "participatory technology development" (Okali, Sumberg, and Farrington 1994). This diversity can be explained, in part, by the fact that the boundaries of the discussion of farmer participatory research are, in effect, defined by the intersection of a number of key themes in current development dis-

Table 2.2.   Suggestions for the Reorientation of Agricultural Research

Suggestion	Reference
If "researchers and funders allocate substantial resources to strengthen the agricultural research capabilities of poorer groups in rural areas of developing countries … [it] would allow them [the farmers] to exert more effective direct control over the content and composition of international and national research and technology promotion systems."	Biggs (1990:1494)
"Unless much more attention is given simultaneously to farmers' technology development in their own right and to outsiders' possibilities in strengthening the experimental capacity of farmers, the gap between farmers and the outside world will remain, and the potential for improvements in agricultural technology will be under-utilised."	Haverkort, van der Kamp, and Waters-Bayer (1991:3)
"Teach them how to improve on these innovations themselves. Through a process of small-scale experimentation, farmers can learn to develop and adapt new technologies. The goals should not be to develop the peoples' agriculture, but to teach them a process by which they can develop their own agriculture."	Bunch (1991:24)
"A central challenge … is for agricultural researchers to appreciate and understand this process of farmer experimentation and to seek ways of articulating on-farm research with farmers' own research projects and modes of inquiry."	Scoones and Thompson (1994b:7)

course. As seen in Figure 2.1, these themes fall into two broad groups: those relating primarily to technical concerns (e.g., second-generation problems with Green Revolution technology, sustainability, marginal areas) and those relating primarily to social and political concerns (e.g., empowerment, the new poverty agenda, the role of government and civil society). These two clusters of themes are joined by the crosscutting themes of participation and local knowledge, thus effectively outlining the arena within which the discussion of farmer participatory research takes place. This figure illustrates the "dual mandate" of farmer participatory research, and it is important to note that its "social development" agenda is considered by many practitioners and academics to be at least as important as its "technology development" agenda.

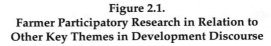

Figure 2.1.
Farmer Participatory Research in Relation to
Other Key Themes in Development Discourse

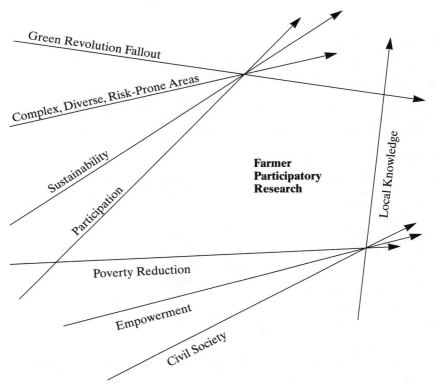

Along similar lines, we previously made a distinction between "research-driven" and "development-driven" farmer participatory research activities (Okali, Sumberg, Farrington 1994). Research-driven farmer participatory research aims primarily to improve the effectiveness of formal agricultural research. Whereas the ultimate goal may be a positive impact on the livelihoods of poor farmers, this goal is approached through the development of new or improved agricultural technology. The work of Sperling and her colleagues in Rwanda (Sperling 1992; Sperling, Loevinsohn, and Ntabomvura 1993; Sperling and Loevinsohn 1993) and of Ashby and her colleagues in Colombia (Ashby, Quiros, and Rivers 1989) are examples of research-driven activities; others include the work of Versteeg and Koudokpon (1993) in Benin; Franzel, Hitimana, and Akyeampong (1995) in Burundi; Potts et al. (1992) in Indonesia; Sumberg and Okali (1988) in Nigeria; and Waibel and Benden (1990) in Thailand. These examples cover crops, livestock, and agroforestry, and the farmers participate in on-station and/or on-farm research. In all cases, greater farmer participation is simply a means to increase the effectiveness of what are, for all intents and purposes, conventional agricultural research programs.

In contrast, development-driven farmer participatory research activities are often associated with projects, community organization efforts, and group-based approaches. Farmer participatory research is seen to contribute to the wider objective of empowerment through the transfer of research skills, increased self-reliance, and the idea that local people can be in a stronger position vis-à-vis formal research and extension institutions.

We also drew attention to the apparent disjunction between the rhetoric that is used to justify and promote farmer participatory research and the reality of the way many development-driven activities are implemented. Thus, although the notion that farmers have a well-established tradition of local experimentation has become almost a mantra among those who promote farmer participatory research, many projects entail training farmers in more organized, systematic, and formal ways of experimentation. We used the decision to strengthen farmers' experimentation through training to distinguish between "less interventionist" (Sumberg 1991; Okali, Sumberg, and Farrington 1994) and "more interventionist" (Bunch 1989, 1991; Gubbels 1988; Ruddell 1995) approaches.

What is particularly striking is that the choice between a more or a less interventionist strategy was seldom based on an analysis of the nature or characteristics of farmers' experiments. Rather, the choice appeared to be based on assumptions that farmers' experiments are either reasonably effective, or that they are haphazard, time-consuming, and prone to misinterpretation. We concluded that unless there was a better understanding of the nature of farmers' experiments, it would be impossible to make rational programming and implementation decisions. Echoing a similar level of disquiet, Bentley (1994:146), in a forthright critique of farmer participatory research, suggests that it has been seriously oversold and that it now needs to be "stripped of its romantic, moralistic tones and judged as any other tool for innovating agricultural technology."

INTEGRATING FARMERS' EXPERIMENTS:
THE SYNERGY HYPOTHESIS

The literature offers two basic justifications for a more active interest in farmers' experiments. The first is that farmers' experimentation offers the potential for the direct empowerment of rural people by increasing their self-reliance and by bolstering their position vis-à-vis the bureaucratic organs of the state, such as formal agricultural research institutions. The second justification is less direct: Integration of farmers' experimentation into formal agricultural research will increase the impact of both the formal and informal research activities, result in better or more appropriate technologies, and eventually contribute to improved rural livelihoods. This dual justification for a concern with farmers' experiments mirrors the split between the social and technology development agendas, a split that is found throughout current development debates. Our concern in this research is primarily with the technology development agenda associated with farmer participatory research.

The case for further integration of informal and formal research is usually developed in terms of creating a more "collegial" relationship between farmers and researchers. In this context, Biggs (1989) described a collegial relationship as "researchers actively encourag[ing] the informal R&D [research and development] system in rural areas." We suggest, however, that this definition is overly restrictive, heavily favors farmers' experiments, and assumes the existence of an "informal R&D system." A broader definition would highlight a relationship based on joint authority, mutual respect, and a commitment to work together as colleagues. Regardless of the exact definition of "collegial," however, the argument is that significant potential benefits, for both farmers and researchers, will flow once more collegial relationships are established (see, e.g., Chambers, Pacey, and Thrupp 1989).

Even for those promoting farmer participatory research simply as a means to improve agricultural technology development and evaluation (leaving aside the suggestion that experimentation per se can empower local people), the prospect of potential synergy (in the sense of the whole being greater than the sum of the parts) following the establishment of more collegial relationships is crucial. We term this *the synergy hypothesis*. Its logic is based on the fact that farmers have an intimate knowledge of their local environment, conditions, problems, priorities, and criteria for evaluation, and they actively engage in experimentation as part of their farming routine. This knowledge, experience, and experimentation is normally out of the reach of "outsiders." At the same time, the results of formal agricultural research are often inaccessible and inappropriate to resource-poor farmers. Using farmers' experiments as the keystone of a collegial relationship between farmers and researchers will yield *significant extra benefits* to both the farmers and the formal research system, benefits that could not be gained through other less collegial modes of interaction or without a focus on farmers' experiments.

Either explicitly or implicitly, the synergy hypothesis is fundamental to most farmer participatory research programs and projects, and it is used to support the argument for a greatly expanded role for farmers and their experiments in formal research. This expanded role includes priority setting, technology development, and program oversight, in addition to their now generally accepted role in problem diagnosis and technology testing. The synergy argument thus implies a radical change in the relationship between farmers and researchers: It is not only a call for farmers to be consulted when problems are being identified, nor does it simply suggest that farmers be given the opportunity to try and to modify the products of research as and when they see fit. Indeed, in taking this argument for an expanded role for farmers to an extreme, Howes (1980) actually called for a breakdown of the division of labor between researchers and farmers. We noted previously, however, that despite this powerful and now widely accepted rhetoric, most participation in farmer participatory research activities continues to take place within formal on-farm trials and thus looks very similar to participation within standard farming systems research (Okali, Sumberg, and Farrington 1994).

In our view, it is essential that the synergy hypothesis be explored and accepted if the radical restructuring of agricultural research that is implied in much of the writing about farmer participatory research is to be justified. The synergy hypothesis raises a number of questions that reflect the contradictions within the discussion and practice of farmer participatory research. The answers to these questions depend directly on the characteristics of farmers' experiments relative to those of formal agricultural experimentation.

We assume, for example, that the net effect of bringing together processes that are similar, that share many essential characteristics, will be additive, meaning that the resulting whole will be *equal to* the sum of the component parts. In this case, the outcome of integrating farmers' experiments and formal experiments might be described not as synergy, but as complementarity. Yet, by bringing together processes that are fundamentally different, and thereby potentially creating a whole new dynamic, the net effect may be both additive and multiplicative, and thus truly synergistic. This assumed relationship between the number of shared characteristics between farmers' experiments and formal experiments and the levels of potential synergy or complementarity is shown graphically in Figure 2.2. Rogers and Shoemaker (1971:14) define "homophily" and "heterophily" as the degree to which pairs of individuals are similar or different in certain respects. The idea is that an increasing degree of heterophily will make communication and understanding among individuals more difficult. In a similar vein, we suggest that a

Figure 2.2.
Potential Synergy in Relation to the Characteristics of
Farmers' Experiments and Formal Experimentation

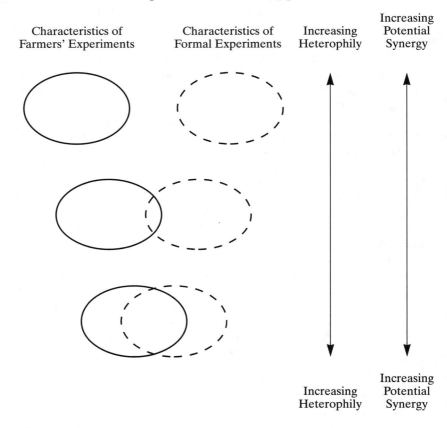

collegial relationship will be most difficult to establish when farmers' experiments and formal experiments have few characteristics in common, exactly the situation in which the greatest potential synergy is also expected.

With these concerns in mind, in the next chapter we review what is known and what has been said about the characteristics of farmers' experimentation and the links between farmers' experimentation and local information networks.

3

Farmers' Experiments: Concepts, Evidence, and Claims

In the discussion that follows, we consider "research" to be "the act of searching (closely or carefully) for or after a specified thing or person; a search or investigation directed to the discovery of some fact by careful consideration or study of a subject" (OED 1992), whereas an "experiment" is "the action of trying anything, or putting it to proof; a test, trial; an expedient or remedy to be tried; a tentative procedure" (OED 1992).

The general notion that farmers are involved in activities that can legitimately be considered as research or experiments is addressed in five separate but related bodies of academic and development literature. These relate to the study of: (1) the prehistoric development of agriculture and the domestication of plants and animals; (2) the development and evolution of agriculture in Europe and North America over the last 300 years; (3) the anthropology and ethnography of "primitive" societies; (4) the diffusion of innovations; and, finally, (5) farmer participatory research. Before reviewing these in more detail, it is important to note that in none of this literature is farmers' experimentation itself the major focus. In addition, this literature reflects a variety of academic disciplines, from history and anthropology to political economy and agronomy, as well as more applied interests, such as rural development and agricultural extension. There is good reason to distinguish among these bodies of literature, but it is also true that there are certain authors who have made significant contributions to more than one.

In reviewing this material, we are concerned, as was Rhoades (1989:4), with discovering "the empirical, as opposed to the romantic or emotional, basis for elevating farmers to an equal partnership in research development."

The first body of literature addresses the historical development of the world's agricultural systems, from the initial domestication of plant and animal species through the development of highly complex systems involving, for example, irrigation, large-scale land management, and the selection of numerous and highly differentiated crop varieties and livestock breeds. Although this literature does not deal directly with the actions of individuals, it is obvious that the evolution from hunter-gatherer to agriculturist, and the subsequent diversity and complexity of the world's agricultural systems, could only have come about by the actions of generations of people who, motivated by necessity, curiosity, or simply coincidence, sought or stumbled upon new ways of doing the numerous individual tasks that result in the production of crops and livestock. We can only surmise that over seasons, decades, generations, and centuries, the knowledge resulting from this "trying" was integrated into, and when accumulated, eventually changed in some radical ways, the techniques, organization, and productivity of farming. While providing perhaps the most persuasive argument for the existence and power of farmers' experimentation and associated networks for diffusion of locally created knowledge, this evidence is circumstantial, and from it we learn little about the processes or mechanics of experimentation. Thus, though this literature alerts us to the historical importance of farmers' experimentation, it adds little to our understanding of its nature or its potential to contribute to more immediate or directed change.

The second body of literature that touches on farmers' experimentation relates to the development of agriculture in temperate regions of Europe and North America over the last 300 years. Throughout the agricultural revolution, there were any number of well-known examples of innovations developed by landowners, farmers, mechanics, and others. Pretty's (1991) review documents a long tradition of farmers' experiments in the United Kingdom, with examples from early in the seventeenth century concerned with soil fertility, crop rotation, sowing techniques, livestock breeding and management, irrigation, drainage, hand tools, and pest control. Although this literature does bring experimentation and change down to the level of individuals, except for the stories of a few prominent individuals, it again provides us with relatively little detail concerning the process or dynamics (i.e., motivation, methods, timescale, outputs, etc.) of farmers' experimentation. Pretty also highlights the importance of farmer-to-farmer transfer of information to the process of agricultural transformation in Britain. It is interesting to note that he places farmers' experimentation *after* the act of adoption: "Once they adopted one of these technologies, [they] experimented to make the changes necessary for its adaptation to local conditions" (p. 139). We will argue later that an important contribution of farmers' experimentation is made *prior* to adoption.

The third body of literature that addresses farmers' experimentation comes from the fields of anthropology and ethnography. Detailed studies of rural communities and economies provide an array of examples of difference in farming systems, techniques, and tools (e.g., de Schlippe

1956; Conklin 1957; Richards 1985, 1986). These differences have been explained, at least in part, in terms of indigenous experimentation and innovation. While other authors highlight the cultural significance of these experiments, Johnson (1972) very effectively draws our attention to their adaptive significance. This literature is central to the current interest in local knowledge, farmers' experiments, and participatory research, and we will make further reference to it as we review the characteristics of farmers' experiments.

The next insight into farmers' experimentation is provided by the vast body of research that began in the early 1940s and that relates to the dynamics and processes of diffusion of agricultural innovations (for a detailed review of much of this material, see Rogers 1983). This work was at the very heart of the discipline of rural sociology in the United States and was (and perhaps still is) the single most important influence on the theory and practice of agricultural extension.

In the minds of many proponents of farmer participatory research, the theory and vocabulary associated with diffusion of innovations research is synonymous with the "transfer of technology" approach, which is itself commonly shunned for being "top-down" and overly dirigiste. It is ironic, therefore, that its theory and models provide a potentially valuable framework for the analysis of farmers' experimentation. It is certainly true that the early research in this tradition was constrained by simplistic notions, including the linear movement of information and technology from central sources, and painted the farmer as a somewhat passive actor who simply made yes or no adoption decisions. However, even the relatively early models of the adoption process generally included a "trial" stage during which farmers were thought to try or experiment with the innovation. And by the late 1970s, the model had been expanded to include both "trialability" ("the degree to which an innovation may be experimented with on a limited basis" [Rogers 1983:231]) and "reinvention" ("the degree to which an innovation is changed or modified by a user in the process of its adoption and implementation after its original development" [Rice and Rogers 1980:500]). These modifications set the stage for the more recent emphasis on adaptation as opposed to adoption and placed the farmer as an active agent in the creation, modification, and diffusion of innovations. Unfortunately, some of these more subtle aspects of diffusion theory have been lost in the implementation of mainstream agricultural extension models, such as the Training and Visit system, and this may partly account for the negative view of diffusion theory in the farmer participatory research literature. Ashcroft and Agunga (1994) recently reviewed diffusion theory in the light of the current interest in participatory development.

The fifth and final body of literature is that which is associated with the more recent interest in farmer participatory research and the broader concern with the effectiveness and efficiency of agricultural research institutions. As indicated in the last chapter, the proposition that farmers do, or are capable of doing, their own experiments is a central tenet of those who promote and implement farmer participatory research. There are, in effect, three major strands to the discussion of farmers' experiments as it relates to farmer participatory research. The first simply supports the proposition that farmers do research as part of their farming activities, citing examples from the anthropological literature referred to above to illustrate differences in farming practice among individuals and sites and over time. Another aspect of this strand highlights a number of examples of more recent innovations that are attributed to farmers, including, for example, the bamboo tube well in Bangladesh (Biggs 1980), diffuse light storage of potatoes in the Andes (Rhoades 1989), and the development and testing of a wide range of crop varieties (Brammer 1980).

The second strand seeks to place farmers' experimentation into context in relation to, for example, formal agricultural research activities and institutions, and agrarian change more broadly (Biggs 1980, 1990). Along these lines, some authors have speculated on the potential impacts on farmers' experiments of closer links with formal research (Van der Ploeg 1990; Richards 1987), while others highlight concerns such as intellectual property rights (Mooney 1993).

The third strand refers to farmers' experiments in relation to alternative models and approaches through which they can be tapped by agricultural research programs (Maurya 1989; Box 1989) and development projects (Haverkort, van der Kamp, and Waters-Bayer 1991). Throughout much of this writing, there is an emphasis on the need for farmers' experiments to be stimulated, strengthened, and supported (e.g., Bunch 1989).

These bodies of literature are reviewed in greater detail in the sections that follow. We are specifically interested in determining what this heterogeneous body of research and writing tells us in relation to the following questions: Do farmers experiment? How common are farmers' experiments? What factors affect the frequency and type of experiments? Who gets involved in these experiments? Is there evidence for the presence of persons who could be described as "experimenting farmers" or "research-minded farmers," and, if so, how are they different from other farmers? What are the characteristics of farmers' experiments; what do they look like? Are they fundamentally different from "scientific experiments" in terms of their methods, logic, meaning, and so on? What are the strengths of farmers' experiments? What are their weaknesses? What role do farmers' experiments play in agricultural change?

EVIDENCE FOR FARMERS' EXPERIMENTS: OUTPUTS AND PROCESSES

The proposition that farmers actively seek out solutions and alternatives, at least partially through activities that can be considered experiments in the sense of "to try," is supported in the literature most commonly through reference to the results or outputs of these activities. For example, the literature that celebrates the complexity and variations of the world's agricultural systems highlights what is assumed to be the output of generations of farmers' experimentation. In much the same way, some authors within the anthropological tradition, and others who cite this literature that identifies differences in agricultural practice among individuals and groups, use these differences to support conclusions about the existence and importance of farmers' experimentation.

Amanor (1994:30), who "explores the adaptive and experimental skills of farmers" and concludes that indeed they "are experimenting with evolving new regenerative technologies," offers a recent example of this approach. To support his conclusion, he cites the change in the cultivation of cowpeas, from a "minor" to a "major rotational" crop, and new bush clearing techniques and fallow management systems, one of which was "methodically created" by a farmer into a "full-fledged agroforestry scheme." Additional examples of what are assumed or asserted to be outputs of farmers' experimentation are the spread of improved varieties of rice and wheat in India and Bangladesh, germination testing of wheat seed on bamboo leaves and adaptation of rice paddle threshers (Biggs 1980), the development of rice varieties, pest control, and fertilizer management strategies (Maurya 1989), intensive multiple cropping systems in Bangladesh (Brammer 1980), the production of new potato varieties in the Andes (Van der Ploeg 1993:213), horticultural practices in Indonesia (Aumeeruddy 1995), and food processing in Nigeria (Waters-Bayer 1988).

However, the vast majority of cited examples of outputs of farmers' experimentation provide little if any indication as to the actual process of experimentation in terms of the motivation, type, or the sequence of activities, duration, sources information or characteristics of the experimenter. Thus, this line of argument essentially confounds the abundant evidence of very rich reserves of "farmer knowledge" with the processes, including experimentation, through which this knowledge was generated. This confusion is important because, as noted by Biggs and Clay

(1981), having detailed knowledge is not the same as being innovative or creative (or, presumably, experimental).

Perhaps more critically, by ignoring the actual processes that together make up the act of experimentation, any sense of the time dimension is lost. We will argue later in the book that the time element is, in fact, one of the most important aspects in the analysis of the potential value of closer interaction between formal research and farmers' experiments.

The other body of evidence for the existence of farmers' experimentation focuses on the processes of ongoing experimentation, and a number of more or less detailed examples have been cited. In making his case for the prevalence and importance of farmers' experimentation, Johnson (1972) cites examples of the small-scale testing of banana varieties and planting regimes from his own research in Brazil. In addition, he trawls the anthropological literature and finds descriptions of the collection, testing, and exchange of new crops and varieties in Ponapean (Bascom 1948), the Philippines (Conklin 1957), Borneo (Freemen 1955), Laos (Izikowitz 1951), and New Guinea (Salisbury 1962; Rappaport 1967). Additional examples of active experimentation by farmers are cited, including the date of planting of tobacco in Puerto Rico (Manners 1956) and the organization of work groups in Zambia. The work of Richards (1985, 1986), which has become so central to the discussion of farmers' experimentation, also falls into this tradition as he describes in some considerable detail the testing of rice germplasm in Sierra Leone. The common elements running through most of these examples are a keen interest on the part of farmers in trying new crops and varieties and the use of small plots for initial testing.

Millar (1993:47) reports that in northern Ghana, farmers' experimentation can be seen as a series of logical stages, including problem identification, testable hypothesis formulation, design, testing, validation, and evaluation and utilization of results. Rhoades and Bebbington (1995:301) describe how farmers in the Peruvian highland zone maintain individual "germplasm banks" made up of "proven" and "reserve" varieties. A few tubers of new varieties are planted "in a kitchen garden or in a short row along a field boundary," and growth and performance are monitored closely throughout the growing season.

Thus, based on descriptions of both outputs and the processes, most observers have concluded that from the Andes to Bangladesh and the Sahel, the propensity of farmers to experiment lies somewhere between ubiquitous (Rhoades and Bebbington 1995:306) and "widespread" (McCorkle and McClure 1995:332). Indeed, Johnson's (1972:156) observation that "experimentation is probably as natural as conformity in traditional communities" sums up a widely stated, general belief that experimentation has been and continues to be a normal, crucial, overlooked, and undervalued part of the practice and the evolution of farming (McCorkle and McClure 1995:332; Biggs 1980; Biggs and Clay 1981; Rogers 1960).

It is in moving beyond this very general level of understanding that the apparent consensus begins to break down. As will become evident in the following sections, there is considerably less agreement in relation to, for example, the actual characteristics of farmers' experiments, the factors affecting the propensity to experiment, and, indeed, the types of people who are engaged in these activities.

It is one thing to establish that farmers do experiments and that over extended periods of time the knowledge so generated fuels the evolution of farming practice and agricultural systems. However, the proposition that farmers' experiments therefore represent an under-valued resource that should be brought closer to the center of formal agricultural research and development activities does not necessarily follow. As has been suggested repeatedly, it is only with a closer examination of the characteristics of these experiments that meaningful decisions relating to the potential value of closer integration can be made (Biggs and Clay 1981; Richards 1985:161; Okali, Sumberg, and Farrington 1994). An understanding of the characteristics and processes of farmer experimentation is also necessary to determine whether or not it is even possible to bring formal and informal research into closer contact.

CHARACTERISTICS OF FARMERS' EXPERIMENTS

The discussion of the characteristics of farmers' experiments must be based on an analysis of the process by which these activities unfold. This analysis will necessarily include a comparison, either implicit or explicit, between "formal" or "scientific" experiments and farmers' experiments. It is, therefore, critical to remember that the practice of formal agricultural research is itself highly variable and often very practically oriented. Much agronomic research, for example, concerns itself with determining optimal dates, spacing, density, rates, and doses, with little if any theoretical underpinning. Care must therefore be taken when comparing formal agricultural experiments and farmers' experiments that we do not adopt an idealized view of "science" that has little in common with the actual practice of agricultural research.

Farmers' experimentation has been described and evaluated in terms of the sequence of steps (i.e., what actually happens), the method, the level of organization and planning, the perspective (i.e., forward- or backward-looking), its accuracy and its efficiency, the measures and criteria used, and the context within which it sits.

We have previously referred to the very formalized sequence of stages that Millar (1993) used to describe farmers' experimentation. Other detailed descriptions of the unfolding of farmers' experiments are given by Richards (1985:98–99), Rhoades (1989:9), Rhoades and Bebbington (1995:301–303), and McCorkle and McClure (1995). At least on the surface, the processes described by these authors are strikingly similar to those of formal experiments. In addition, there is an array of examples of individual experiments that were conceived and implemented by farmers. There is almost unanimous acceptance of the proposition that farmers' experiments take place, at least initially, on small plots (Rogers 1960; Lionberger 1960:23; Biggs 1980; Rhoades 1989; Pottier 1994). Johnson (1972:157) refers to this as a "slow cautions experimental approach," while Rhoades talks of "step-wise" adoption. Only Connell (1991:218) highlights farmers' "tendency to use the chosen technology over the entire field," thus making comparisons difficult.

The examples range from relatively simple tests of a new crop variety to more complex, multiyear evaluations of fallow systems. Many experiments are reported to include side-by-side comparisons of varieties or treatments, which Biggs (1980) refers to as "yes/no trials." Rhoades and Bebbington (1995) and McCorkle and McClure (1995) indicate that farmers' experiments are sometimes repeated over several years with the effect of adding an element of replication to what is otherwise a potentially confounded comparison. In relation to farmers' tests of potato varieties, Prain, Uribe, and Scheidegger (1992:56) note that "in farmer experimentation, 'temporal' replication is considered a more relevant way of dealing with variation."

Although there are clearly similarities between farmers' experimentation and formal agricultural experimentation (Rhoades 1989:8; Potts et al. 1992; Scoones and Thompson 1994a:29), Cornwall, Gujit, and Welbourn (1994:100) warn against drawing "erroneous parallels" between the two. In fact, many authors prefer to highlight the differences, with farmers' experiments being described as unsystematic, chaotic, and unorganized (Gubbels 1988:12; Millar 1993:48) and as "trial and error" (Biggs 1980; Millar 1993; Abedin and Haque 1991:162). Going further, Stolzenbach (1994:155) questions the value of the notion of identifiable experiments as opposed to the "continuous and innovative element of the craft of farming." More specifically, Rhoades and Bebbington (1995:307), for example, indicate that "unlike scientists, farmers show limited concern with statistical proof," while Rhoades (1989:9) suggests that farmers have very specific goals in mind and have little interest in experimentation for its own sake. Biggs and Clay (1981) indicate that major limitations of farmers' informal research are the facts that it cannot address long-term problems and that it is restricted by locally available genetic resources and unsystematic transfers of genetic materials. In addition, they claim that it is not "forward looking," meaning that it is of limited value in helping farmers to anticipate opportunities.

Moving beyond these operational differences, much is made of the notion that farmers' experimentation is deeply embedded in local social, cultural, and economic systems. Thus, Amanor (1993a:41) sees an intimate connection between experimentation and processes of labor, which results in "a holistic approach which is concerned with synergism, systems dynamics, and the interrelationship of the energy cycles within the farm environment." Along similar lines, Van der Ploeg (1993:210) uses the term *"art de la localité"* to refer to the process by which local knowledge is generated through the interaction of labor and the local environment. Others, such as Sharland (1991:148) and Pottier (1994:86), see fundamental and culturally determined differences in terms of both the concepts underlying agricultural practice (integrated and holistic) and approaches to evaluation (not necessarily based on measurement). In reference to his work in the Dominican Republic, Box (1989:66) observed that the trials of researchers and farmers were "aimed at different objectives, used different methods, and produced results which are not verifiable by the other party." He concluded that the logic was so different that it was "hard [for researchers and farmers] to come to terms." This proposition of fundamental differences between farmers' experiments and formal experiments is developed by Van der Ploeg (1989, 1993), who argues that the respective "research traditions" (i.e., scientific on the one hand and *art de la localité* on the other) are so different in inner logic, scope, and dynamics that there may not even be agreement on what constitutes an experiment or an innovation. Indeed, he suggests that from the point of view of formal research, farmers' *art de la localité* is a "superfluous or even counter-productive element" (1993:220). As will be seen in later sections of this chapter, the argument that farmers' experiments and formal experiments are different is used to highlight the potential danger to farmers that this difference will be lost if their experimental activities are integrated into, or "co-opted by" the formal research system.

One final characteristic of farmers' experimentation that has received some attention is the degree to which it is or is not a public affair. Some authors describe farmers' experimentation as an essentially private matter, at least until the experiment is finished (Gubbels 1988; Pottier 1994:84; Millar 1993:48). Indeed, it has been suggested that those innovations that are new and exciting may be least likely to move into the public domain (Cornwall, Gujit, and Welbourn 1994:106, citing Fairhead 1990). Similarly, Dunkel (1985) suggests that at least in Rwanda, poor extension services and a restricted flow of information tend to give farming techniques the character of "family secrets," which will severely limit the potential impact of farmers' experimentation. In contrast, Winarto (1994:152) paints a somewhat different picture when describing how new knowledge from ongoing experiments with the White Rice Stem Borer is being widely shared.

It is assumed in much of the current literature that farmers' experimentation will need to be a much more public affair if it is to play a larger development role. Gubbels (1988:12), for example, makes it clear that the need is to move from individual experimentation that is neither "systematic" nor "organised" toward a "concerted, organised community effort to determine priorities, analyse problems, seek solutions, and take actions." This same view is evident in Vel, Velduizin, and Petch (1991:152). Others, such as Connell (1991:220), claim that it is necessary to stimulate farmers' "latent ability to experiment," while Bunch (1991:35) seeks to move beyond simply trying new things by teaching farmers "a method of village research."

TYPOLOGIES OF FARMERS' EXPERIMENTS

Farmers' experiments have been described and categorized in a variety of ways, based on their objectives, motivation, degree of planning, and so forth. In their discussion of reinvention, for example, Rice and Rogers (1980:508) describe two main categories: "planned" and "reactive" (i.e., because of unexpected or unsatisfactory consequences of the original innovation). They also suggest that for the particular studies they reviewed, "management" reinvention tended to be planned, "technical" reinvention tended to be reactive, and "operational" reinvention was split between the two types.

Rhoades and Bebbington (1995:299) classified farmers' experiments with potatoes in terms of the experimenter's motivation and identified curiosity experiments, problem-solving experiments, and adaptation experiments. Based on work in northern Ghana, Millar (1993:47; 1994:161) added peer-pressure experiments to this list. Stolzenbach (1994:156), using work by Schön (1983), analyzed examples of farmers' experiments from Mali and suggested that they can be exploratory, move-testing, and hypothesis-testing at the same time. Since reflection and action overlap in farmers' experiments, and since experiments are an integral part of the larger, dynamic "performance" of farming (Richards 1989), Stolzenbach suggested that attempts to categorize (or clearly define or identify) them may be of limited interest and value. This conclusion would probably find favor with those who argue that farmers' experiments, by virtue of their embeddedness in local culture and context, are fundamentally different from formal experiments (e.g., Box 1989; Cornwall, Gujit, and Welbourn 1994).

In an earlier review of farmer participatory research, we followed Rice and Rogers (1980) in suggesting a framework to differentiate among farmers' experiments based on the degree of active planning or control (Okali, Sumberg, and Farrington 1994). Thus, a "proactive" experiment, on the one hand, demands a conscious move to create specific conditions or treatments and subsequent observations that are more or less systematic. A "reactive" experiment, on the other hand, has no systematically preset objectives, treatments or observation criteria, and "much of the experimental process is probably left to circumstances or hazard which determines the context and particular conditions in which observation can create new understanding"(p. 130).

Apart from the diffusion of innovations research, the discussion of types of farmers' experiments is based on the analysis of a very limited number of selected examples. There is no basis, therefore, on which to compare the relative value of these typologies, or to make any measured judgments as to the relative frequency of different types of experiments or the factors associated with the different types. It seems likely, however, that a solid basis on which such judgments can be made will be essential if fundamental shifts in the relationship between farmers' experiments and formal research are to take place.

FACTORS INFLUENCING FARMERS' EXPERIMENTATION

It seems clear that the discussion now needs to move beyond the simple idea that farmers do experiments if it is to be of any real value to those who plan and implement agricultural research and development activities. As will become evident in the paragraphs that follow, some authors have attempted to develop this general proposition by highlighting a number of factors that appear to or might possibly affect particular farmers' interest in, or ability to carry out, certain types of experiments. The specific factors that are identified in the literature include the degree of poverty, spatial isolation, and contact with extension services, markets, and experts; diversification within the farming system; environmental degradation; and familiarity with the specific environment; as well as gender, social, and moral relations more generally. These factors may affect differences in the level or characteristics of farmers' experimentation both within and among sites, and over time. It is important to note that there are few studies that address this question either directly or indirectly. Perhaps the most prominent body of literature comes from the diffusion of innovations school and is focused on the socioeconomic and personality characteristics associated with differences in innovative behavior, but these more personal, within-site differences are dealt with in more detail in the next section. The value of this literature is also somewhat limited for the purpose of understanding farmers' experimentation in that the critical indicator is usually the "adoption" of a given technology, whereas individual acts of farmer experimentation may or may not result in a change that can be clearly identified as adoption.

The question of the impact of poverty on farmers' experimentation has been addressed in terms of both motivation and ability. Swift (1979:42) and Gamser and Appleton (1995) indicate that poverty itself may be an important element of the motivation to experiment, as poor people are "forced" to experiment in order to survive. Indeed, Gamser and Appleton assert that "increased poverty ensures increased growth of peoples' technologies"(p. 345). Conversely, there are those including Amanor (1994) and Winarto (1994) who maintain that poverty, drudgery, and risk-averse behavior hamper the ability of farmers to experiment. Their ability is lessened not so much by a lack of motivation or interest, but by a reduced capability to follow through with the experiment and to carry the risks associated with unproved practices. Poverty and a lack of control over productive resources may well make it more difficult to repeat an experiment over a series of years and thus confirm observations through replication over time. For example, it is interesting to note that in the Andean communities studied by Van der Ploeg (1993:223), rich farmers did not change completely to the new potato varieties. In contrast, farmers who had land but lacked the means to cultivate it changed to the new varieties more rapidly and completely as they needed the credit that was offered as part of an "integrated rural development" program, and could not gain access to it without planting the new varieties. According to Amanor (1994), this poverty-related constraint to experimentation represents a real loss, as "promising lines of research never get fully developed" but "remain as bright but passing ideas."

The question of the risk inherent in experimentation is critical and was addressed directly and succinctly by Johnson (1972:151):

> What must be made clear . . . is that experimentation and risk are separate matters; it is possible to experiment at low cost and low risk. It is true that traditional agriculturists are cautious (conservative) in the face of innovations; it is not true that they refuse to try them out. Somehow these two notions have become intertwined and confused.

Johnson's view is certainly supported by most of the descriptions of the scale of farmers' experiments: Even accepting that informal experimentation is not without cost (Biggs and Clay 1981), the livelihood of even the poorest farmer is unlikely to be put at greater risk through planting a new variety, for example, in a small garden plot. The ability of poor farmers to plan and execute more complex experiments may well be limited, but to date there is little evidence that complexity is a common characteristic of the experimentation of any small-scale farmers.

Poverty can be reflected in the degree of diversification within a farming system, in geographical isolation, and in a lack of access to formal extension services. All of these factors have been proposed as affecting farmers' experimentation. Thus, Rhoades and Bebbington (1995:300) hypothesize that high levels of diversification and poor extension services may make farmers more interested in experimenting. This would lead one to expect a negative relationship between the level of farmers' experimentation and the degree of specialization within the agricultural economy (this question is addressed in Chapter 7). This line of argument is extended by Amanor (1994), who suggests that the "commodity-focused research tradition" essentially creates a monopoly over technology generation and thus saps the initiative of farmers to experiment for themselves. Indeed, Van der Ploeg (1993:221) argues that modern agricultural science makes farmers "invisible men," an idea extended by Pretty (1991), who notes that by repeatedly associating the beginning of agricultural experimentation with the establishment of formal experiment stations, the contribution of years of sophisticated and valuable experiments by farmers is masked.

In contrast, Chambers, Pacey, and Thrupp (1989:49) propose that contact with the market and specialists actually stimulates experimentation; and Vel, Velduizen, and Petch (1991:152) see small farmers in isolated areas as having a limited capacity to "analyse their situation critically and think of it objectively as something that can be altered through their own action." Thus, consciousness raising and increased self-confidence are prerequisites for meaningful farmers' experimentation. One aspect of physical isolation is population density, which Richards (1985:86) saw as having no obvious effects on the inventiveness of farmers in West Africa.

Another aspect of poverty is environmental degradation. In one of the few field studies that purports to address the factors affecting the level of experimentation, Amanor (1994:29) concludes that among his study sites in Brong Ahafo, Ghana, the loss of the forest frontier makes innovation an imperative, and thus "the more degraded environments are found to be hotbeds of experimentation." It is, however, interesting to note that this contrasts with his other general conclusion that experimentation is restricted, rather than stimulated, by poverty and resource scarcity. Along similar lines, Farrington and Martin (1988:29) suggest that indigenous knowledge, and presumably the local experimentation that supports it, may break down when people are faced with an environmental crisis.

It is not unreasonable to assume that the level or type of experimentation in a particular locality will reflect the importance of factors affecting stability, change, or new opportunities. Bentley (1994), building on the work of Goldman (1991), indicates that farmers may be experimenting more now in "the face of a rapidly changing environment and the availability of new biological and chemical products." Similarly, Rhoades and Bebbington (1995:306) see some evidence that points to an interaction between the context and the type of experiment, with adaptation experiments being more common in areas of "considerable change" and curiosity and problem-solving experiments dominant in more stable environments. A special case in terms of rapidity and degree of change is that of migrants to new areas and new agroclimatic conditions. Rhoades and Bebbington (1995:302) address this situation relative to the movement of migrants from the highlands to the lower areas in Peru and conclude that experimentation "is one of the fundamental strategies involved in the settlers' attempt to learn about and control their environment." They note that in an area with a long history of in-migration, established migrants have given up trying to grow potatoes, but recent migrants continue to try anyway. They deduce from this that "innocence of possibilities is one of the positive points favouring creative experimentation." This situation in Peru seems to be in stark contrast to an area of recent resettlement in Ethiopia, where few if any examples of farmers' experimentation could be identified (Sandford 1990).

A range of sociocultural and familial factors have also been associated with the propensity of farmers to experiment. Drawing on their work among the Susu of Sierra Leone, Longley and Richards (1993:51) conclude that experimentation takes place within society-specific moral orders, and, consequently, an individual's attitudes toward experimentation may reflect his or her place in society. It follows that an understanding of farmers' experimentation will necessitate an analysis of "the sociocultural context in which local agricultural knowledge is generated" (ibid.:56). Indeed, some have even suggested that it is the farmer's "cosmovision" that creates the context for experimentation (PRATEC 1991; Millar 1993).

Gender relations within the family have also been highlighted as influencing experimentation and innovation (Chambers, Pacey, and Thrupp 1989:47). In relation to farmers' experimentation in Rwanda, Pottier (1994:84, citing Dunkel 1985) suggests that a limited flow of new genetic material and management techniques into the family locus acts to restrict experimentation. Presumably, this is one of the concerns Gubbels (1988) is trying to address in Burkina Faso through village meetings. Rhoades (1989:6) indicates that social factors are important not only in terms of the level of experimentation and who participates, but also in terms of their direct effect on the objectives of the experimentation.

In discussing the phenomenon of "reinvention," which would seem to be the more general case of farmers' experimentation, Rice and Rogers (1980:501) suggest that it is more likely to occur when technologies are complex and irreversible. The other important characteristic of a technology that may affect the degree of experimentation or reinvention is its "trialability" (Rogers 1983:231), or the ease with which its elements can be disentangled and tested individually and on a small scale. Reinvention is described as being necessary in situations where a technology must be made to work in a wide variety of situations, which might lead us to expect more intense farmers' experimentation in the so-called complex, diverse, risk-prone areas. Reinvention is also supposed to be of particular importance to "early adopters." It should be noted that both Biggs and Clay (1981) and Richards (1985:14) suggest that farmers' experimentation is something more than a simple process of adaptation: It is a continuous, dynamic, and innovative process. But this contrast between adaptation on the one hand and innovation on the other would appear to be a false dichotomy, in that whether farmers' experimentation is focused on internally or externally generated ideas or technology, its processes and outputs, and those of reinvention more generally, can be both adaptive and innovative.

WHO EXPERIMENTS?

In addition to the factors discussed in the preceding section, there is a need to determine if there are within-site differences in the innovative or experimental behavior of individuals, and, if so, what factors might help account for them. There would appear to be two basic positions: The first is that everybody and anybody experiments, and thus the interpersonal differences are, in fact, minimal. The second position is that there are important differences in the innovative behavior of individuals that can be accounted for by personal, social, cultural, and economic factors.

There are many claims to the effect that all or many farmers experiment, but there is relatively little direct evidence concerning the frequency or distribution of experimentation within farming populations. Thus, Johnson's (1972) proposition that experimentation is probably as common as conformity is supported by Rhoades and Bebbington (1995:296), who reported the results of a survey showing that 90 percent of settled farmers in the upper Chanchamayo area of Peru were "avid experimenters." McCorkle and McClure (1995), based on work in Niger, also conclude that experimentation is widespread.

On the other hand, there is at least one very substantial body of research that points to significant differences in innovative behavior among farmers in a wide range of contexts. The diffusion literature further distinguishes between innovators and "early adopters": Although early adopters are also more educated and have higher social status than average farmers, as well as being slightly younger than average farmers, neighbors are more likely to approach early adopters than innovators for advice. Furthermore, Rogers suggests that the trialability of an idea or technology is more important to innovators and early adopters than to later adopters.

The propositions that innovative capacity is unevenly distributed within and between communities and that the ability of individuals to generate new knowledge varies significantly (Chambers, Pacey, and Thrupp 1989:37) are certainly in line with this analysis. The more recent literature on the participation of farmers in research adds a number of other names to describe farmers who do experiments: "rural innovators" (Biggs 1980), "innovative farmers" (Biggs and Clay 1981), "experimenting farmers" (Box 1989), "research-minded farmers" (Biggs 1990), "real innovators" (Bentley 1994), and "farmer scientists" (Winarto 1994). The definitions of some of these categories are shown in Table 3.1.

The definitions given in Table 3.1 clearly imply that these individuals are different from other farmers in that they experiment more often or more effectively, they are in contact with other experimenting farmers, or they are widely known for their experimentation. However, apart from the diffusion of innovations literature, there appears to be no clear research basis for these categories and their associated assumptions. Rather, they are based on a limited number of examples—of farmers' experiments and experimenting farmers—that are seldom placed in any larger context. In fact, there are a number of important contradictions within the discussion of farmers who experiment or innovate. For example, in relation to the selection of participants for on-farm trials, Merrill-Sands et al. (1992:129) suggest that researchers "need research-minded farmers who are good collaborators in order to ensure the quality of experimental research results." On the other hand, Rogers and Shoemaker (1971:183) describe innovators as desiring "the hazardous, the rash, the daring, and the risky," which are not the kind of desires we might associate with a reliable research collaborator.

We have previously suggested that different kinds of experiments might be associated with different types of individuals (Okali, Sumberg, and Farrington 1994). Assuming that reactive (i.e., essentially unplanned) experiments would be more common than proactive experiments, we suggested that although anyone and everyone might do (or have no

Table 3.1. Categories of Farmers

Category	Definition	Reference
Innovators	The first 2.5% of individuals to adopt a new idea	Rogers (1983)
Early adopters	The next 13% of individuals to adopt a new idea	Rogers (1983)
Innovative farmers	"Always experimenting with new technologies or changed technical possibilities"	Biggs and Clay (1981:327)
Real innovators	"Will go on to develop new, appropriate technologies or devise effective modifications of existing practices"	Bentley (1994:147)
Farmer scientists	"Participants who regularly [make] observations and [discuss] their findings with others"	Winarto (1994:152)
Research-minded farmers	"Farmers, village artisans, etc., who are always experimenting in one way or another and are involved in informal R&D and diffusion activities"	Biggs (1990:1485)
Experimenting farmers	Those respected for their knowledge about a particular crop; part of an invisible network of those who experiment	Box (1989)

choice but to do) reactive experiments, a smaller number of individuals, presumably among those identified in Table 3.1, are likely to be associated with proactive experiments. Building on the descriptions of innovators provided by the diffusion of innovations literature, one can identify personal characteristics, including age, gender, and level of education, that might be associated with different types of experiments. We will return to these issues in Chapter 7.

LINKS WITH FORMAL RESEARCH: FARMERS' EXPERIMENTS IN A LARGER CONTEXT

Much of the literature reviewed to this point concerns itself essentially with farmers' experimentation as an activity undertaken by individuals. However, some authors consider these individual experimenters as members of larger networks. For example, Biggs and Clay (1981) describe an "informal research and development system," while Rhoades and Bebbington (1995:298) refer to a "people's scientific community." Biggs and Clay identify four important differences between formal and informal research systems: In formal systems, the generators and the users of technology are different; formal systems have institutional memory, links outside the local environment, and the capacity to plan for the future; formal systems are more affected by government policies; and formal communications are of greater importance in formal research systems. Nevertheless, the formal and informal systems are described as parallel and analogous (Rhoades and Bebbington 1995:298; Biggs and Clay 1981).

Unfortunately, the basis for the jump from examples of individual experiments and experimenters to the conception of either an informal research and development "system" or a people's scientific "community" is tenuous at best. The image implied by these conceptions is of a relatively high level of coordination or linkage among those involved in informal experimentation. To our knowledge, however, there is at present little data to support this notion (apart, perhaps, from Millar's [1993:47] description of "peer-pressure experimentation"). Implicit in the notion of an informal research and development system is the assumption that some individuals, along the lines of the research-minded farmers described in the previous section, do experiments more frequently or of a different type than other farmers. Again, apart from the data on innovators and early adopters provided by diffusion of innovations research, the distinction between research-minded farmers and other farmers has not yet been established through empirical research.

We find this leap from the proposition that farmers do experiments to the notion of an informal research "system" both curious and critical. On the one hand, it adds further legitimacy to farmers' experimentation by moving it beyond the realm of individual, isolated acts. By placing farmers' experiments in their own institutional framework, the notion of an informal research system also sets the stage for a more serious discussion of the ways and means to closer integration of formal and informal research activities. The informal research system becomes a natural "counterpart" to formal research. Rhoades and Bebbington (1995:98), for example, suggest that "dovetailing indigenous farmers' experiments with scientists' experiments is one option to improve the generation and transfer of appropriate technologies for traditional agriculture," and many others have come to a similar conclusion. The recent interest in farmer participatory research is one response, and the interventionist approaches advocated by Bunch (1989, 1991) and Gubbels (1993) can be seen as attempts to actively create structures to which outsiders can more easily relate. In other words, the notion of an informal research system that is somehow the logical counterpart to the formal system helps set the stage for efforts to make informal experimentation more formal.

The enthusiasm for bringing farmers' experimentation into closer contact with formal agricultural research is certainly not shared by all. Both Richards (1987) and Van der Ploeg (1990, 1993) suggest that farmers' research traditions and activities have been, are being, or can be weakened or co-opted through certain kinds of relations with formal research. Farrington and Martin (1988) are moving in a similar direction in indicating that indigenous technical knowledge (and, presumably, farmers' experimentation) may break down when people are faced with "external interventions." In addition, concern has been expressed about the commoditization of knowledge and the growing hegemony of formal research and agribusiness interests (e.g., Amanor 1994:15; Mooney 1993) and thus a loss of self-reliance on the part of small-scale farmers.

Once again, however, there is little evidence in the literature either to support these general propositions or to enable the formulation of more specific hypotheses. Are farmers in commercial, highly mechanized areas more or less "experimental" than small-scale farmers in marginal areas? Do closer links with formal research foster or inhibit the ability of farmers to address their problems and concerns through reinvention? Does the notion of an "informal research system" accurately describe the relations among experimenting farmers? These are some of the questions that will be addressed in later sections of this book.

FARMERS' EXPERIMENTS AND
LOCAL COMMUNICATION SYSTEMS

If farmers' experiments are about the creation of new knowledge, then their significance is in part dependent on the subsequent movement of the information on which that knowledge is based. However, the local informal systems through which individuals, groups, and communities "acquire, encode and transmit *their* knowledge" (McCorkle 1989; emphasis added) have received relatively little research attention, particularly in comparison to that given to formal extension systems, despite the suggestion that local information systems are a logical starting point for developing formal agricultural extension systems (ibid.).

As is the case for farmers' experiments, insights about local systems for the exchange of agricultural information are scattered throughout several bodies of literature. These include cultural anthropology and early network studies (Mitchell 1969); sociology and the network approach (Boissevain and Mitchell 1973); rural sociology and studies of the diffusion of innovations (Rogers 1983); education; and systems of learning and knowledge creation. Discussions in each of these areas contribute to the understanding of local processes of communication and how these can be explored through field studies. One of the first challenges is to define such local systems, a task that some have suggested is as problematic as defining "local" or "indigenous knowledge" (Mundy and Crompton 1995).

Clearly, our own interest in local communication is very specific: to explore its relationship to farmers' experiments. Thus during the period of our fieldwork presented in this book, we were interested in assessing whether people are aware of what other people are trying and the significance of this awareness, as well as assessing the importance of different sources of information for the experimentation process itself. Ultimately, we hoped to assess the potential for integrating farmers' experiments and formal research through stronger links between local and formal communication systems.

The question of the nature of the links between farmers' experimentation and local communication is simply not addressed in the literature. In a recent review of the anthropological literature on local communication networks, Hall (1994) looked specifically for information relating to the characteristics of individuals who are central to these networks, the types of groups involved, how groups emerge and change, and the value of information exchanged. She concludes that although considerable attention is given to the general topics of exchange and communications, specific reference is seldom made to movement of technical information relating to agriculture. Two possible reasons are suggested: the general lack of interest in the subject on the part of anthropologists, and the possibility that the technical aspects of agriculture are not widely discussed within communities.

However, technical and nontechnical information, and new or modified agricultural practices, clearly move within and between communities. There have been numerous studies, dating back to the 1940s, of the movement of information about agricultural innovations that have been specifically introduced into a community by an outside agency such as an extension service or a commercial firm. In a more recent study of the diffusion of "true potato seed" technology in Indonesia, Chilver and Suherman (1994:302) conclude that "it is the motives and attitudes of key individuals involved in informal diffusion, over and above the institutional system of which they are a part, that are crucial in determining who stands to benefit from new technologies." They also note that the dissemination of technology is driven by social obligation and altruism, on the one hand, and commercial motives, on the other, with the latter becoming more important as the innovation becomes established.

Nevertheless, the movement of locally generated information and innovations has received little attention. Hall (1994), citing work by Netting (1968), Robertson (1978), and Aboyade (1987), notes that "farmers ask questions about things they have seen even as they gossip and use humour, sarcasm and ridicule to mock each others' farms and skills." Farmers also closely observe the actions of other farmers and copy what they think would be useful for themselves (Robertson 1978; Batterbury 1993; Anthony and Ochendu 1970; Bowen 1993). They visit extensively (Evans-Pritchard 1940; Batterbury 1993; Anthony and Ochendu 1970), and they swap, and on occasion even steal, seed and other material (Netting 1968; Hill 1963).

Some authors have emphasized the wider social and cultural significance of these exchanges (Boster 1986; Pottier 1994; Longley and Richards 1993), and there is evidence that technical information and seed do not necessarily move smoothly and equitably among community members (e.g., Sperling and Loevinsohn 1993; Longley and Richards 1993). However, to date, there has been little attempt to determine the significance of these observations in terms of the interface between local and nonlocal information and exchange networks.

With regard to the potential for stronger links between formal and informal communication systems, we have earlier noted the suggestion by Rogers (1993) that the role of formal extension is to enhance local systems. Certainly, formal extension, even in its most traditional forms, attempts to make use of informal communication networks. The work on the diffusion of innovations highlighted the role of the "opinion leader" within local communities (Rogers 1983). And today, "contact farmers," whose task it is to draw the attention of others to new technology, are a central feature of the T&V extension system. With the objective of broadening both participation and impact, one common adaptation of the T&V system is to work with local groups rather than individual contact farmers (Bagchee 1994).

Central to any discussion of information exchange are the ideas of trust and gossip. Within communities characterized by close ties of kinship and marriage, the fact that there is both a positive and a negative side to information sharing has certainly been acknowledged. Paine (1967), for example, reviews earlier material that states that negative community interpretations of gossip are designed to assert the unity of the group that may otherwise be damaged. He also offers an alternative analysis that presents gossip as simply another form of informal communication that is purposeful and managed and can be used to protect individual interests. The implications of the roles of trust and gossip in local information exchange must be borne in mind by those who propose to use local networks for development purposes. Trust and gossip also featured in our own field research.

SUMMARY

This chapter has highlighted widespread acceptance of the proposition that at least some farmers engage in activities that can be considered experiments. Based on fieldwork from several academic traditions and an array of examples, we can say that the idea that farmers engage in the development, testing, and manipulation of ideas and technology relating to farming is very well established. While there has been considerable interest in farmers' experimentation and participatory research since the early 1980s, the importance of farmers' experimentation was recognized by observers of British agriculture as early as the eighteenth century (Pretty 1991). And for well over fifty years, the importance of farmers' experimentation has been recognized explicitly by mainstream academics concerned with agricultural change and agricultural extension.

Nevertheless, among those interested in participatory research, who either forgot or rejected the frameworks provided by the large body of theory and field data associated with diffusion of innovations research, there is now relatively little agreement as to either the personal or the contextual factors affecting an individual farmer's propensity to experiment. Is experimentation stimulated or hampered by poverty, by rapid environmental, social, or economic change, by access to markets, by contact with formal research and extension?

This review highlighted a number of hypotheses, some of which are contradictory, that are present in the literature. These can be summarized as follows:

1. Farmers engage in activities than can be labeled "experiments."
2. Farmers' experiments:
   2.1. Concern both technical and nontechnical issues;
   2.2. Share some characteristics with formal research;

2.3. Are socially and culturally embedded; they are based on fundamentally different frameworks compared to formal research;
2.4. Are more than simply adaptive;
2.5. Are not forward looking;
2.6. Are haphazard, trial-and-error, inefficient, and have methodological limitations;
3. Socioeconomic, personal, and agro-ecological factors increase (+) and decrease (−) the propensity to experiment, both within and between sites:
3.1. Between sites: environmental degradation (+); diversification (+); isolation (+/−); access to research and information (−); influence of commodity-oriented research (−); stability (−);
3.2. Within sites: poverty (+/−); degree of "research-mindedness"(+).
4. Farmers' experiments need to be, and can be, strengthened or supported.
5. Strengthening farmers' experiments empowers farmers.
6. Farmers' experiments represent an untapped development resource.
7. Closer integration of farmers' experiments and formal research will result in synergistic benefits.

This list illustrates the variety and apparently contradictory nature of many of the claims and hypotheses in the recent literature. For most of these claims, there is little beyond circumstantial and anecdotal evidence on which to build a case one way or the other. In fact, a surprisingly large part of the recent literature on farmers' experimentation and participatory research revolves around a relatively limited number of examples and cases. To date, there has been little attempt to put these into context or to look systematically at the characteristics of the process of experimentation itself. This lack of empirical evidence accounts in part, we suggest, for the gap between the rhetoric of the larger debate about farmers' experimentation and participation, on the one hand, and the reality of many contemporary farmer participatory research projects, on the other (Okali, Sumberg, and Farrington 1994).

The literature offers little basis with which to address the two central questions identified earlier: What benefits can be expected from closer integration of farmers' experiments and formal research? Does farmers' experimentation need to be "strengthened" before these benefits can be realized? In the chapters that follow, we will use the results of field research from both Africa and the United Kingdom to attempt to shed some light on these important questions.

4

Research Methods

Our analysis of farmers' experiments and their links with local information networks draws on two main sources of data. The first is fieldwork
undertaken in Kenya, Zimbabwe, and Ghana between November 1994
and February 1995. The second source is a series of recent studies in the
East Anglia region of the United Kingdom (Rijal, Fitzgibbon, and Smith
1994; Lyon 1994; Knight 1995; Carr 1996). In the sections that follow, we
introduce the site selection and data collection methods.

SITE SELECTION

In previous chapters, we highlighted dominant themes and approaches
in agricultural research and extension in sub-Saharan Africa. In addition, the review in Chapter 3 demonstrated the need to consider context,
at national and local levels, when seeking to understand variation in
farmers' experimentation. A number of key factors were identified as
potentially having an impact on farmers' experimentation. These include
level of poverty, spatial isolation and population density, access to markets and levels of commercialization, type of farming system, environmental degradation and familiarity with the environment, social relations of production and norms in relation to new ideas and practices, and
contact with formal research and extension services.

Thus, the choice of Kenya, Zimbabwe, and Ghana as research sites
reflects the possibility of identifying loci that exhibit significant variation
in some of these factors. The choice of these countries was also influenced by their relative political stability, our own previous experience
and contacts, a desire for a relatively wide geographical spread, and the
explicit interests of our funders. We then sought to identify sites that en-
compass significant diversity in agro-ecology, population density, settlement pattern, level of commercialization, and degree of accessibility.
This process of site selection was also influenced by our collaborating institutions: the International Centre for Insect Physiology and Ecology
(ICIPE) in Kenya, the Intermediate Technology Development Group

(ITDG) in Zimbabwe, the Cocoa Research Institute of Ghana (CRIG), and the Wenchi Farm Institute (WFI) and the Wenchi Farming Systems and Training Project (WFSTP) in Ghana.

The combination of these institutional links and our interest in the specific site characteristics listed above led us toward western Kenya, southern Zimbabwe, and eastern and central Ghana. In selecting specific settlements, we were concerned to avoid unnecessary overlap with ongoing research or development activities. Only in Zimbabwe was our collaborating institution working directly with the community we studied. The sites finally selected were Kangare village in Nyanza Province, Kenya; VIDCO D, Ward 21 in Masvingo Province, Zimbabwe; Boma and Suminakese villages in Eastern Region, Ghana; and Pamdu and Awisa villages in Brong Ahafo Region, Ghana. These sites will be described in detail in the following chapter.

RESEARCH METHODS

Kangare in western Kenya was the first field site, and it served primarily to develop and test the research methods. This included identifying categories of people to be interviewed; addressing possible problems with group, rather than individual, interviews; determining terminology to be used; and assessing the feasibility of visiting fields and livestock for more focused discussions. The work necessarily developed in an iterative manner. We finally selected Kangare partly for its size and settlement pattern. It is a small, compact settlement, which allowed us to return more than once to some households and, by covering almost all the households, to try to address issues of community networking. Some discussions were held with groups of people, while others were held with individuals. Following some initial lengthy discussions, one older man and his son drew a map of the village and provided some background information about each household. Interviews were eventually conducted in fourteen households and included a number of visits and discussions in fields or areas where livestock were kept.

We finalized our sampling criteria following the work in Kangare. Our intention was to take a purposeful sample of local residents. In addition, we hoped to interview key individuals who were identified as such by other respondents and to follow up on specific activities such as tomato production, which was identified as particularly interesting by our informants. For this, we used some key personal and socioeconomic criteria that are likely to affect the dimensions of one's engagement in farming: gender, age, marital and family status, and the presence of outside income. Practically, except in the case of Ward 21 in Chivi, where

ITDG had previously assembled detailed individual and household information, it was only possible to sample by age and sex. In Chivi, it was possible to select a random sample from the following categories: (1) male-headed, household head's primary occupation is farming; (2) male-headed, household head's primary occupation is not farming, but he is normally resident; (3) female-headed, household head is widowed, divorced, or unmarried; (4) female-headed, spouse normally resident elsewhere.

This type of information was not available for any sites in Ghana, and even the latest census data were not accessible. Moreover, at the sites in Eastern Region, it was particularly difficult to determine marital status. Therefore, at Suminakese and Boma, approximately equal numbers of males and females within three age classes (15–35, 36–55, and 56+ years) were interviewed. At these sites, no attempt was made to prepare a map or to put together a list of households. As in Kenya, we had no introduction from an outside agency, and we were especially concerned with not appearing to be closely associated with official cocoa activities, which involve farm surveys and payments for cutting out of trees. In Boma, especially since the village was waiting to appoint a new chief, we were very aware of the need to avoid the appearance of any political alliances as well. At Awisa and Pamdu in Brong Ahafo Region, village maps were drawn by local people, who also provided some indication of the socioeconomic situation of the individual households. These maps, combined with a listing of names garnered through initial interviews, provided a rough sample frame from which individuals were selected based on gender, age, marital status, and level of engagement in farming.

Interviews were conducted either in local language through an interpreter or, when possible, in English, and they took place either within the fields or at the homesteads of the respondents. At one site, Suminakese, many of the fields are at a distance from the village, and farmers were usually willing to visit only if they were also harvesting crops. Most interviews were with individuals; however, in a number of cases, spouses and/or children were present. Interviews lasted on average between thirty and sixty minutes. A minimum of socioeconomic information was compiled for each individual respondent, including the level of engagement in farming, gender, age, marital status, origin, community status, level of formal education, off-farm employment history, previous contact with agricultural extension and development projects, primary and secondary occupations, and whether or not remittances were regularly received. In the light of time limitations and our desire to raise neither suspicions nor expectations, we did not attempt to estimate or rank the relative wealth of the respondents.

In general, the interviews consisted of a personal introduction by the researcher, followed by a brief introduction to the topic. Following this, one of two general approaches was taken. The first, an indirect approach, built on the framework proposed by Okali, Sumberg, and Farrington (1994:132), which involves discussion focused on current farming activities (i.e., those undertaken the day of the interview or the day before) with the objective of identifying differences between "normal" and "actual" practice. In other words, was there anything peculiar, different, or special about how the respondent plowed (planted, weeded, etc.) the day of the interview compared to how he or she did so a year prior? The second approach was more direct and consisted of a straightforward explanation of our interest in farmers' search for new ideas and techniques, followed by a request for any examples they could provide from their own experience. In some cases, the respondents were asked to react to a simple story that contrasted internally and externally driven agricultural change.

We were particularly concerned with avoiding possible bias resulting from the use of terminology such as "experimentation," "research," "innovation," and "extension"—all of which are associated with "formal" systems and may be automatically associated in peoples' perception with particular types of activities. The problem was to identify something (i.e., farmers' experiments) that we expected to be both different from the formal notion of research and experimentation and at the same time largely embedded within the daily routine of farming. Thus, at each site, the first task was to agree on a vocabulary: At all sites, examples of experimentation were usually associated with, or indicated by, local phrases for "to try" and "I tried."

Following the work of Okali, Sumberg, and Farrington (1994:130), we considered two conditions necessary for an activity to be considered an "experiment": the creation or initial observation of conditions or treatments, and the observation or monitoring of subsequent results or effects. When differences between normal and actual practice were identified, or when examples of "tests" were reported, these two conditions were evaluated. If satisfied, the example was added to a catalog of farmers' experimentation, and the circumstances surrounding the activity or change were explored as far as possible. Of particular interest were questions of motivation, the source of the idea, the method used to test it, and any involvement of other people. Examples of experimentation were subsequently classified according to (1) the subject of the example, (2) the motivation behind the experiment, (3) the method used, (4) the source of the idea, and (5) the real or potential outcome of the experiment (see Table 4.1).

Table 4.1. Classification of Examples of Farmers' Experimentation

Aspect	Category
Topic of the experiment	new crop new variety spacing/density fertilizer/soil fertility marketing arrangements land preparation/seeding method control of crop pests pruning water harvesting/erosion time of planting crop combination irrigation labor arrangements crop storage manure management control of livestock pests weed control variety by tillage method crop residue management crop by site
Motivation	proactive; inquisitive; tries new things reactive; improvisational no information
Method used	no obvious "control"; an all-at-once change somewhat formalized testing; split field or other immediate "control" no information

Source of the idea or
  technology
	copying/trying something that was observed/
	    suggested
	trying something that has been or is being actively
	    promoted
	own idea
	no information

Actual/potential outcome	novel technique, crop, process, or organization rediscovery of widely known technique; minor modification to known technique major modification no information

Using these methods, examples of farmers' experiments were collected from many respondents. However, at all sites there were some respondents, and at one site as much as 73 percent of respondents, who did not report any experiments. We do not know whether the fact that particular individuals did not report an example of an experiment means that they *never* experiment. If an individual's engagement in experimentation is sporadic or associated with particular stages in the life cycle, it is certainly conceivable that memory recall of specific past experiments may be partial at best. We are also very well aware that some individuals, and often particularly women, are not forthcoming in a formal interview situation. We deliberately tried to take steps to overcome this widely recognized problem of muteness, by, for example, conducting interviews in private, visiting the respondent's field or plot, and generally following an informal and flexible protocol. Nevertheless, this problem cannot be disregarded.

With respect to our interest in sources of agricultural information, local networks, and the identity and characteristics of "innovators," "local experts," and "dynamic" and "successful" farmers, these issues were explored through direct questions. Thus farmers were asked, for example, where the particular idea for an experiment originated. Otherwise, they were asked to name individuals associated with the introduction of specific ideas and practices, and to name sources to which they would go if they were in need of information on specific subjects (e.g., diseases on tomato plants).

In total, 189 interviews were conducted (see Table 4.2). A limited number of group interviews were also carried out, such as an interview we conducted in Chivi with seventeen farmers from a neighboring settlement. In addition, our fieldwork included discussions with other farmers, government officials, agricultural researchers, extension personnel, and agricultural merchants, as well as participation in events such as the 1994 Farmers' Day Competition in Tafo District, Ghana.

CHARACTERISTICS OF RESPONDENTS

Overall, 56 percent of the respondents were male, ranging from 64 percent in Kangare to 41 percent in Chivi (see Appendix Table 1). The modal age was 36–55 years (43 percent), with the remaining respondents almost equally distributed between the younger and older age classes. Respondents in Kenya tended to be older, while those in Brong Ahafo were younger than the sample as a whole, which reflects the attention given in this area to the particular growth industry of tomato

Table 4.2.   Distribution of Interviews by Site (in number)

Country	Site	Women	Men	Total
Kenya	Kangare	5	9	14
Zimbabwe	Ward 21	29	20	49
Ghana: Eastern				
Region	Suminakese	15	16	31
	Boma	14	28	42
	Subtotal	29	44	73
Ghana:				
Brong Ahafo	Pamdu	6	16	22
	Awisa	14	17	31
	Subtotal	20	33	53
Total		83	106	189

production. In terms of formal education, 53 percent reported 0–4 years, ranging from 27 percent in Kangare to 74 percent in Eastern Region. The proportion of respondents with more than nine years of formal education ranged from 1 percent in Eastern Region to 46 percent in Brong Ahafo, Ghana.

Across all sites, 70 percent of the respondents were married and 24 percent widowed or divorced. The highest percentage of respondents who had never been married was in Brong Ahafo (12 percent) and the lowest in Kangare (0 percent). The vast majority of respondents reported farming as their primary occupation (ranging from 93 percent in Brong Ahafo to 71 percent in Kangare) and had a resident spouse (ranging from 90 percent in Kangare to 73 percent in Chivi). Thirty-five percent of respondents reported receiving some remittances (ranging from 56 percent in Kenya to 17 percent in Brong Ahafo). Approximately half of all respondents had worked outside the immediate area of the village, ranging from 88 percent in Kangare to 47 percent in Brong Ahafo, and 20 percent had either worked for the agricultural extension services or had some direct contact with a development project (other than ITDG in the case of Zimbabwe).

Compared to men, women respondents generally had less formal education (45 percent of men with less than five years compared to 63 percent of women), were more likely to be widowed or divorced (men 5 percent, women 49 percent), less likely to have a resident spouse (men 96 percent, women 59 percent), more likely to receive remittances (men 28 percent, women 43 percent), and less likely either to have worked outside the area (men 72 percent, women 28 percent) or to have worked with the agriculture department or a development project (men 28 percent, women 10 percent) (see Appendix Table 2).

DATA FROM EAST ANGLIA

Postgraduate students in the 1994 Farming Systems Analysis course of the School of Development Studies, University of East Anglia, used a framework similar to that described by Okali, Sumberg, and Farrington (1994) to identify examples of farmers' experiments. They interviewed pig, sugar beet, and vegetable producers (Rijal, Fitzgibbon, and Smith 1994). Building directly on this experience, Lyon (1994) subsequently interviewed eighteen pig and arable farmers in Norfolk and Suffolk Counties during the summer of 1994. Farms ranged in size from 100 to 900 acres, and the age of the farmers ranged from thirty to seventy years. Using an "open-ended" approach, Lyon sought to examine how farmers do experiments and the potential for integrating these activities with formal agricultural research.

Taking a different approach, Knight (1995) used the Ideas Competition of the Royal Norfolk Agricultural Show to examine farmers' experimentation and innovation in the area of farm machinery. The competition began in 1948 and has been organized by the Norfolk Farm Machinery Club (NORMAC) since 1971. NORMAC was founded in 1946 to provide a forum for information exchange and the improvement of the standard of education and competence among machinery operators. The club now has 649 members divided into twelve local centers, each with its own committee. A newsletter is distributed to all members in addition to some 600 other individuals and companies in the agricultural machinery trade. NORMAC sponsors events such as the Ideas Competition that bring farmers and machinery manufacturers together, as well as machinery demonstration and tractor handling competitions.

Prizes and medals are awarded to entrants in the competition (farmers, farm employees, machinery manufacturers) on the basis of the novelty and value of their "idea" (which must be a working example). The competition is normally judged by officials from institutions such as the Agricultural Development and Advisory Service or the National Institute of Agricultural Engineers.

Knight created a database that included information about essentially all entries and entrants in the Ideas Competition from 1948 to 1994. In addition, he interviewed entrants and visitors to the 1994 Ideas Competition at the 1994 Royal Norfolk Show. Knight argued that the 460 entries in the resulting database provide a unique opportunity to examine the characteristics and implications of farmers' experimentation and innovation within a very particular context. With Knight's database in hand, Carr (1996) subsequently used a postal survey to gather information from eighteen farmers who entered the Ideas Competition between 1988 and 1994. This survey covered the process through which the "idea" developed, subsequent use of the idea by the farmer and his or her neighbors, and any attempts to patent and commercialize the idea. In addition, eight of these farmers were subsequently interviewed in greater depth.

5

Research Sites

COUNTRY CONTEXTS

Both the population distribution and the character of agriculture in Kenya and Zimbabwe reflect the effects of the colonial, white settler, and plantation economy, and present agro-industrial development. In both countries, historically, the state was used to safeguard these interests, systematically constraining indigenous agricultural development by reserving high-potential agricultural areas and markets for settlers.

For example, one of the most important and persistent impacts of the colonial and pre-independence experience in Zimbabwe relates to the legal classification and distribution of land. The net result of many decades of policy and programs that were designed to serve the interests of the minority, white commercial farmers is that the vast majority of black farmers find themselves on relatively small plots in low-potential, communal areas (Mehretu 1994:57). In 1982, it was estimated that 73 percent of the rural population lived in communal areas (ibid.:56, citing Central Statistics Office 1991), and almost three-quarters of all communal land falls within zones with an annual rainfall of less than 650 millimeters. A desire for more equitable access to better-quality land was an important factor motivating the fifteen-year struggle for independence. Since independence in 1980, government programs have resettled 52,000 families, or 16 percent of the population estimated to need land in 1980 (Bratton 1994). The success of these resettlement programs has been mixed, and even today the land issue remains at the top of the political agenda (Roth 1994).

Much of the literature covering southern and eastern Africa emphasizes the impact of the settler economy as a whole, including agriculture, mining, and urban development, on patterns of rural livelihoods. Kenya in particular is noted for the extent to which women form the core of the country's smallholders and farm labor force. Many rural areas are characterized by more or less permanent male out-migration, with women remaining in the villages producing the basic foodstuffs required for their own and their children's well-being. In this sense, Kenya typifies the female-farming systems that were first highlighted by Boserup in 1970 and that are now being actively addressed by government and development organizations in Kenya and throughout much of Africa (Boserup 1970; Saito, Mekonnen, and Spurling 1994).

In Ghana, the nature of the agricultural sector today is also, in part, a result of the colonial interest in agricultural exports. However, in contrast to Kenya and Zimbabwe, Ghana experienced less discontinuity with precolonial rights of access to land, and the cocoa economy was and still is based on the initiatives of local, usually small-scale producers. More generally, agriculture in Ghana is dominated by smallholders, although there are some large farms and plantations for rubber, oil palm and coconut, and, to a lesser extent, rice, maize, and pineapples (Ministry of Agriculture 1991). In spite of the government's historical interest in mechanized farming, for the most part, agriculture in Ghana remains largely unmechanized. And although women are also heavily involved in agricultural production and may dominate the population in certain rural areas, they are also renowned for their economic independence and even control some sectors of the agricultural economy.

Although there are important differences in the history, structure, and performance of the agricultural sectors of Ghana, Kenya, and Zimbabwe, the present policies and programs of these countries broadly reflect those seen throughout the continent. Thus, the structural adjustment themes of market liberalization, privatization, subsidy reduction, and export-led growth feature more or less strongly in each of the three countries. In Kenya, for example, Cleaver and Donovan (1995) note the slow pace of economic reform and numerous setbacks, in spite of the fact that Kenya was one of the first countries in sub-Saharan Africa engaged by the World Bank in structural adjustment lending. As an example of the slow pace of reform, they cite the liberalization of the maize market, which occurred only in 1994.

In addition to the land reform and resettlement programs referred to previously, agricultural policy in Zimbabwe since independence has focused on marketing, credit provision, and reform of the research and extension services (Rukuni 1994:31). With the provision of grain and cotton marketing outlets in communal areas, some farmers have taken advantage of a particular set of circumstances to significantly increase their production and sales of key crops such as maize and cotton. Rukuni and Eicher (1994) call this Zimbabwe's "second agricultural revolution" and identify five "prime movers" that have made it possible: new technology; human capital; sustained growth of biological capital; improved performance of marketing, credit, research, and extension institutions; and favorable economic policy and political support. Adding a note of caution, however, a number of writers indicate that this second agricultural revolution has not resulted in the elimination of widespread rural poverty or food insecurity (Stack 1994; Mehretu 1994).

Although agriculture is no longer the largest contributor to Ghana's gross domestic product, agricultural development policy has had considerable impact on both rural and national development. The resolve with which Ghana has implemented its Economic Recovery Programme since 1983 is considered by many to be exemplary. So, for example, of twenty-nine countries reviewed by Cleaver and Donovan (1995), between 1981 and 1991 Ghana increased the real producer price for exports more than any other. While cocoa, which is traditionally the most important agricultural export, continues to receive considerable policy attention and investment, in recent years there has been a strong interest in alternative export crops and food crops. However, despite this, in the early 1990s, the agricultural growth rate was "in the vicinity of a disappointing 2 percent per annum" (ibid.:29).

The standard of the national research systems supporting the agriculture sectors in all three countries was defined as "minimally acceptable" (ibid.:37). Still, all three are generally recognized as having a relatively strong and well-established tradition of agricultural research focused on the key commercial crops. In Kenya and Zimbabwe, the research systems were developed to serve the settler economy, although, as noted above, smallholders in Zimbabwe are now reaping some of the benefits of these previous investments.

In Ghana, research previously focused principally on cocoa, but there are now a large number of different state organizations involved in agricultural research, and there is a clear division between the research and extension services for cocoa and for other crops. Although the Ministry of Food and Agriculture is responsible for the development and implementation of general agricultural policy, cocoa remains the responsibility of the Cocoa Marketing Board.

Largely under the auspices of the World Bank, both Kenya and Ghana are undertaking a reorganization of their national research and extension systems, although in the case of Ghana, cocoa research and extension are excluded. As in thirty other African countries, the T&V extension system has been introduced, and a "unified" extension service, at least for food crops, is being promoted. It is noteworthy that in 1982 Kenya was the first African country to initiate a pilot T&V extension program, and the extension service is noted as being particularly good in working with female farmers (Saito, Mekonnen, and Spurling 1994).

While extension systems within sub-Saharan Africa generally have a poor reputation, they have been assessed as performing better than the research services. In Kenya, extension is assumed to be at least partly responsible for the comparatively high average growth rate within the agricultural sector since 1967 (between 3.9 and 4.4 percent)(Cleaver and Donovan 1995). As a result of this growth, almost all food consumed in Kenya is produced domestically, and the agricultural sector is also the main source of raw materials for a growing agro-industrial base (Saito, Mekonnen, and Spurling 1994).

Zimbabwe also experimented with the T&V system, but it was rejected as a national model because it was too expensive (Rukuni 1994:32; Pazvakavambwa 1994:107). Other group-based extension models are now being developed to replace the Master Farmer Training Scheme and the Farmers Clubs, which were central elements of the government's agricultural extension strategy in the communal areas. The first Master Farmer certificate was awarded in 1934, and throughout Zimbabwe 40,000 farmers were awarded Master Farmer certificates before 1980; an additional 80,000 have been awarded since independence (Mutimba 1994). Historically, the Farmers Clubs brought together those individuals who had received training as Master Farmers; fee-paying members of a Farmers Club are exempted from sales tax on agricultural materials.

In each of the three countries, therefore, the agricultural economy has two distinct sides: One consists of specialist enterprises that are likely to be larger, commercially oriented, capital and technology intensive, and well serviced, while the other is dominated by smallholders who use draft animals or hoes and market crops when they can and who combine a number of on- and off-farm activities. The problems of smallholders, who are the majority in all three countries, have become central to the agricultural research and extension services only very recently. Ironically, these states have now started a process of disengagement, and private-sector involvement in agriculture is being encouraged, even in sectors such as cocoa in Ghana and maize in Zimbabwe, which play central roles in the national economies.

STUDY SITES

Our research on farmers' experiments took place at six sites within the three countries (Map 5.1), and the general characteristics of these sites are shown in Table 5.1. The individual sites are described in more detail in the sections that follow.

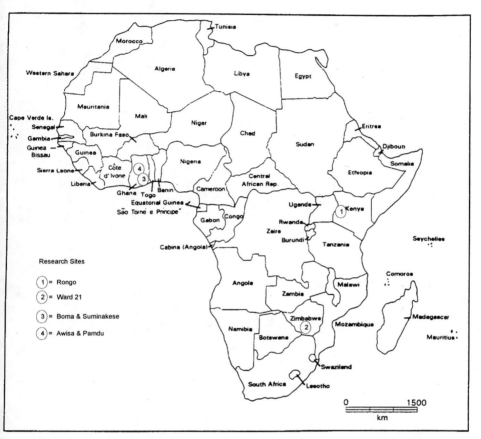

Source: Developed from Figure 1.2 in Robert Stock, *Africa South of the Sahara: A Geographical Interpretation* (New York: Guilford Press, 1995) with permission of the publisher.

Table 5.1. Characteristics of Research Sites

Characteristic	Kenya: Kangare	Ghana: Eastern Region (Boma/Suminakese)	Ghana: Brong Ahafo Region (Awisa/Pamdu)	Zimbabwe (Ward 21)
Settlement pattern	semidispersed	nucleated	nucleated	semidispersed
Distance to major center	3 km	13 km/18 km	3 km/30 km	64 km
Condition of access road	excellent	poor/very poor	excellent	excellent
Degree of agricultural commercialization	low	high	high	medium
Land sales	yes	minimal	minimal	no
Importance of remittances	high	high	unknown	high
Technical service presence	high	high/medium	low	medium
Farmers' groups presence	low	none	medium	high
NGO presence	high	low	medium-high	high
Annual rainfall	1,200 mm	1,300 mm	1,400 km	600 mm
Major cereal crops	maize; sorghum	maize	maize	maize; sorghum; millet
Major cash crops	maize; sugarcane	cocoa; cassava; vegetables	vegetables; maize	maize; groundnuts
Perennial crops	sugarcane; napier grass	cocoa; plantain; citrus	none	none
Vegetative crops	cassava; sweet potatoes; sugarcane	cassava; yam; plantain; onion	cassava; yam	none
Routine seed purchase	medium	low	low	high
Livestock	cattle; goats	goats; sheep	cattle; goats; sheep	cattle; goats; donkeys
Animal traction	medium	none	none	high

Fieldwork Site 1: Western Kenya: Kangare Village,
Kabuoro Sublocation, Kamagambo Location,
Central Division, Migori District, Nyanza Province

From an agro-ecological point of view, western Kenya is highly variable. Jaetzold and Schmidt (1982) identified six agro-ecological zones ranging from the high-altitude and high-rainfall coffee-tea zone to the lower-altitude and lower-rainfall livestock-millet zone (Table 5.2).

Kangare village lies in the transition zone between lower, drier areas farther to the west, and higher, wetter zones to the east. With an average annual rainfall of approximately 1,400 millimeters, much of this transitional area has two assured growing seasons and consequently is considered to have relatively high agricultural potential. Nevertheless, a long tradition of male out-migration and the availability of attractive economic alternatives mean that agriculture is not at the top of the local development agenda.

Kangare is a relatively small Luo village with approximately twenty households, all but two of which belong to one "family" related through a common grandfather. The people of Kangare are closely related to the residents of five nearby villages. Population density in the area is in the range of 300 persons per square kilometer and appears to be increasing (Jaetzold and Schmidt 1982:147). Other indicators of the intensity of land use include well-defined field boundaries, a degree of land fragmentation, and the registration and sale of land.

Kangare is served by good transport infrastructure and has good market access: It is situated 3 kilometers south of Rongo town on the main tarmac road between Rongo and Homa Bay, 30 miles to the south. It also has direct access to the towns of Migori and Kisii. A selection of farm and veterinary inputs is available for purchase in Rongo town, and residents of Kangare have ready access to the offices of the Department of Agriculture and the Veterinary Department in Rongo. In addition to one agricultural extension officer who has responsibility for the Kangare area, a retired agricultural extension officer lives and farms in the village itself.

Maize is the staple grain crop, but many other crops are also grown. Local residents report that millet, groundnuts, sesame, and cowpeas are declining in importance, while crops such as "hard" sugarcane, tomato and other vegetables are now being grown commercially. In recent years, there has been increased interest in short growth cycle sorghum and napier grass (*Pennisetum purpureum*), the latter being tied to the promotion of milk production under zero grazing regimes. Consequently, many of our discussions focused on these crops. There is

Table 5.2.   Agro-ecological Zones of Western Kenya

Zone	Name	Altitude (m)	Rainfall (mm)
UM1	Coffee-tea	1,500–1,600	1,600–1,800
UM2	Coffee	1,450–1,700	1,400–1,600
LM1	Lower-mid sugarcane	1,300–1,500	1,600–1,800
LM2	Marginal sugarcane	1,300–1,500	1,300–1,700
LM4	Marginal cotton	1,140–1,350	800–1,300
LM5	Lower midland livestock-millet	1,135–1,300	700–900

Source: Adapted from Jaetzold and Schmidt (1982:130).

also a sugar out-grower scheme in the area and a small amount of tobacco production.

Animal traction is used for land preparation, but there appears to be
only limited purchase of inputs for crop production such as improved
maize seed. Most fields are farmed by individuals, although some joint
farming between spouses was reported. The most commonly cited agricultural needs and problems were short-cycle sorghum seed, animal pest
control, declining soil fertility, bean pests, and the parasitic weed striga.

As in much of western Kenya, off-farm employment is an important
element of the livelihood strategies of the residents of Kangare. There is
much temporary and long-term migration of males in search of work.
Men from Kangare were reported to work at a range of jobs, including
tractor driver on a sugar estate near Kisumu, laboratory technician at a
marine research institute in Mombasa, and school inspector in Kisii.
Most of the married women with absent husbands are resident in households headed by men.

Although women work family fields, they are also involved in
both the planning and the day-to-day management of their own fields.
Male farmers appeared to be either older individuals or younger men
who have yet been unable to secure off-farm employment: There
would appear to be relatively few men or women for whom farming is
a "chosen profession." Nevertheless, this is a relatively high-income
area: A 1991 survey indicated that annual household expenditure in
the area averaged 152 percent of that of South Nyanza Province as a
whole (Anon. 1991).

Residents refer to three formal groups in the village relating directly
to agriculture: The St. Joseph's Group is focused on dairy production
and has received dairy cattle and other inputs from Heifer International,

an international NGO; the Women's Group was also formed primarily as a vehicle for the distribution of free dairy cattle, but it is not presently active (membership was never limited to women); and the Rongo Coop Dairy Society provides the official market for fresh milk.

Perhaps the most significant development in the village in recent years has been the introduction of grade milk cows kept under zero grazing. With the activities of various local and international programs and groups, the first animals were introduced in 1981, and there are now approximately 1,268 grade cows under zero grazing with 720 farmers in Rongo Division. The Veterinary Department estimates that about half of these were gifts associated with both government and nongovernment programs, although farmers are now reportedly purchasing grade cattle with their own resources. Feed largely consists of napier grass, which has now assumed an important role in the cropping pattern: Some farmers were reported to be producing napier grass as a cash crop.

Fieldwork Site 2: VIDCO D, Ward 21,
Chivi District, Masvingo Province, South-Central Zimbabwe

Zimbabwe is divided into five distinct agro-ecological zones, with average annual rainfall ranging from over 1,000 millimeters in Zone I to less than 450 millimeters in Zone V (Chasi and Shamudzarira 1992). Chivi District falls within Natural Regions 4 and 5 and has between 300 and 650 millimeters of "erratic and very low" or "unreliable" rainfall. Extensive livestock or semi-intensive livestock production with drought-resistant crops in the wetter areas are recommended (ibid.). Within the district, there are essentially two soil types, one being sandy (sandveld) and the other having a higher clay content and better water-holding properties (clayveld). Wilson (1990) was able to relate the differential soil-moisture relations of these two soils to differences in the degree of agro-ecological productivity, food supply, and consumption. Differential access to these soils, both between and within communities, thus has important implications for food and livelihood security. Short- and long-term migration by men in search of work are well-established strategies for dealing with the precarious agroclimatic conditions and land scarcity.

Overall, Ward 21 in Chivi represents a difficult, relatively low-potential site. The low and erratic rainfall constrains all agricultural activities and encourages investment off-farm whenever an opportunity arises. Recent droughts have resulted in the loss of livestock, a significant demechanization of crop production, and an increase in artisanal gold mining. Food production is highly variable, and both off-farm income and emergency food distributions make important contributions to livelihood security. Despite the gains made since independence, poverty continues to be important in Ward 21: Livelihood diversification is a key element of survival strategies in an area that is only marginal for dryland crop production.

Administratively, communal areas of Zimbabwe are divided into wards, which usually comprise approximately 600 families, and Village Development Committees (the VIDCO is a concept introduced in 1984 by the postindependence government and usually regroups several "traditional villages" or kraals totaling approximately 100 families). Ward 21 lies in the southern half of Chivi District with a population of 6,000–7,000 people and approximately 600 households. The ward covers six VIDCOs, with VIDCO D including four kraals: Dzviti, Margwiro, Mapundu, and Gatawa, each having a leader or kraal head (*sabhuku*). The population density of the district is approximately fifty-four persons per square kilometer, and the total population grew from an estimated 137,000 in 1982 to 171,753 in 1994.

Using seventy-four years of rainfall data from Chivi, Balderrama et al. (1987:16) calculated that there was sufficient rainfall for a good harvest in 0.6 years out of 5. It is reported that as a result of the 1991 drought, farmers in Chivi lost up to 80 percent of their cattle and 50 percent of their donkeys (K. Murwira, personal communication). The district experienced severe drought in 1992–1993, and as of December 1994 when we were carrying out our fieldwork, the government was still distributing food and seeds in response to the drought.

Like Kangare, the VIDCOs within Ward 21 are serviced by good transport infrastructure: They are located off the main Masvingo-to-Beitbridge tarmac road and are 60 kilometers from Chivi town, the district's administrative headquarters. They are also served by a network of relatively well-maintained dirt roads. While the settlement pattern is dispersed, there are several small shops scattered throughout Ward 21, which, in addition to staple domestic items, also stock some agricultural items such as seed maize and plow parts. Most major purchases have to be made in Chivi town or Masvingo, which can be reached by a daily bus service.

The growing season is between November and March, and the major crop is maize, which is produced largely from short-cycle hybrid varieties. Treated seed of the most common variety (R201) is available in local shops. Other crops include finger millet (*rapoko*), bullrush millet (*mhunga*), sorghum, groundnuts, bambara groundnuts, and cotton. The average size of farm holding in the Shinde area of southern Chivi District was estimated recently by Corbett (1994) to be 1.92 hectares, while Balderrama et al. (1987) cited estimates of between 2.15 and 3.23 hectares.

As in Kangare, there is a well-established tradition of off-farm employment in the district. One study found that 13 percent and 48 percent of household income was derived from off-farm income and remittances, respectively (GFA 1986). In their own survey, Balderrama et al. (ibid.:87) found that 40 percent of households had income from the formal sector, 65 percent from migratory seasonal employment, 42 percent from local wage labor, and 80 percent from self-employed activities. The centers of male employment most commonly cited by our informants were mines and sugarcane estates, although the drought has also disrupted the sugar industry, and many men have returned to the VIDCO. Most men interviewed have worked outside the area at some point: Some who were absent over a period of up to twenty-five years eventually returned to farm. Others migrate for only short periods and have essentially remained full-time farmers.

Local opportunities for income generation off-farm include crafts such as sewing and weaving, small-scale gold panning, and fishing. During the period of our research, many men and women were panning for gold three or four afternoons per week, as there were already warning signs that the rains might not develop "normally."

The prevalence of off-farm employment for men has important implications for the overall management of the household and the agricultural activities of its members. Thus, while Corbett (1994:22) found that only 13 percent of households in the Shindi area of Chivi District were headed by a divorced or widowed female, "many more women than this assume day to day responsibility for their households in practice." Although it is generally assumed that female-headed and female-managed households are at greater risk than male-headed households, there is some evidence to the contrary (Stack and Chopak 1990). For Chivi, Balderrama et al. (1987) found that households with off-farm employment tended to own more cattle and were more likely to own an oxen-drawn cart and to have larger families.

There are fourteen Farmers Clubs with one Area Committee within the ward, and until recently, membership in Farmers Clubs in VIDCO D was restricted to Master Farmers. However, the clubs now welcome a wider membership, including women, serve as a major channel of agricultural information to the community, and also organize rotating labor groups. It is interesting that many individuals report themselves as being members of a Farmers Club and participate in activities but do not regularly pay the membership fee and so do not receive the government subsidies.

There has been a wide range of development initiatives within Chivi District since the mid-1980s, including income-generating programs, water provision of various kinds, small livestock and grazing schemes, and school construction (Murwira n.d.). These activities have been sponsored through government and nongovernmental organizations. The

ITDG has been active in the ward since 1989, with an emphasis on participatory research around the issue of soil and water management (Murwira et al. 1995). The ITDG has worked closely with Agritex, the government extension service, and the Farmers Clubs, and has promoted the more open attitude toward Farmers Club membership.

Fieldwork Sites 3 and 4: Eastern Region, Ghana:
Boma Village, Akyem Abuakwa Traditional Area,
Suhum District; Suminakese Village, Mpraeso District

The last part of our fieldwork took place at four sites in Ghana, two largely in the forest zone and two within the forest-savanna transition. All sites share a history of cocoa production: The villages of Boma and Suminakese in Eastern Region lie within the oldest cocoa area in Ghana, while Pamdu and Awisa in Brong Ahafo Region are at the northern periphery of the cocoa belt. Farmers at all four sites are also engaged to a greater or lesser extent in commercial vegetable production.

Eastern Region lies largely within the rain forest zone, although toward the north it moves into guinea savanna grassland. Total annual rainfall averages between 1,400–1,800 millimeters and is distributed in a bimodal pattern, with peaks between May–June and September–October. Eastern Region is also one of the more densely populated areas of the country, with 105 persons per square kilometer (Ministry of Agriculture 1991), and is settled largely by the Akan people. It is important to note, however, that immigrant labor from the north and east has played a major role in the development of cocoa production in this area. And like most cocoa production areas in Ghana, Eastern Region is highly politicized.

In spite of the long history of swollen shoot disease of cocoa, and the widespread destruction of trees throughout Ghana by fires in 1982–1983, cocoa remains an important crop in Eastern Region. Cocoa production continues to be used as an indicator of wealth for those dependent on farm incomes (Baah et al. 1994); however, not everyone has a cocoa farm, and this is especially true of women and young men. Also, the decline of production since the 1960s has encouraged new interest in food crop production in these old cocoa areas. Because Eastern Region was the center of cocoa production, there are numerous agricultural research and extension stations and substations there, including CRIG and the main training center of the Cocoa Services Division, the extension arm for the sector.

In Eastern Region, nonfarm income is very important, and the region must be seen as comparatively wealthy. It has a long history of investment in education, much of it through income from cocoa, and its proximity to employment opportunities on the coast means that rural livelihoods are, in general, highly diversified.

Overall, Boma and Suminakese are reasonably well-endowed sites from the point of view of rainfall, and their agriculture has long been dominated by tree crop production. However, the long-term negative effects of cocoa pests and diseases, and the proximity to the off-farm opportunities offered by Accra and other nearby towns, have sapped the agricultural sector of its dynamism, and there appears to be little new investment in agriculturally related activities. In these essential ways, the two sites are representative of large areas of the West African forest zone.

Boma is a comparatively small, nucleated village with 127 houses situated within the Akyem Abuakwa Traditional Area. A total population of 851 persons was recorded in 1984 (396 male and 455 females), and 1,063 persons was projected for 1994. Although this can no longer be described as an area of primary forest, timber extraction continues and is much in evidence. Most residents work within agriculture, although there are also a number of tailors, two shopkeepers, and several school-teachers. Apart from church activities and drinking (there are two bars), the village offers few diversions.

Situated only 4 kilometers from a major tarmac road that connects the northern areas of Ghana with Accra, the village gives the impression of a quiet and neglected backwater hemmed in between the more dynamic and thriving centers of Kwabeng (4 kilometers east) and Awinare (5 kilometers west). With no major market and only one local primary/middle school, transport, other than timber trucks, passes through Boma to Awinare only irregularly. Because the village lies within Eastern Region, however, specifically within the Akyem Abuakwa Traditional Area with its long tradition of education and rural-urban population movement, Boma is neither socially nor economically isolated. For example, all our informants had relatives, usually offspring, living outside the village, and many had relatives living and working in Accra itself. A recent decision to give out building lots in the center of the village has stimulated much new construction, although it was pointed out that the financing for this was coming from outside the village.

The area is also rich in minerals, and men are involved in gold and diamond mining especially now that activity within the mining sector has been liberalized. The common surface mining methods require little capital investment, which is said to account for an influx of young men into the area. A number of Boma men have worked with "company" diamond fields situated near Kwabeng and Kade.

In addition to cocoa, traditional forest zone crops including maize, plantain, cocoyam, and cassava dominate agriculture in Boma. There is much intercropping of food crops, although a few farmers monocrop maize, cassava, and, to a lesser extent, plantain. A few young men are moving into commercial vegetable production, particularly tomatoes and eggplants. Small ruminants, the most common livestock species, are generally tethered or penned following a government health ruling that is apparently enforced through routine, weekly visits and financial penalties for defaulters. There are no cattle in the area.

The village has a small marketplace where a few women gather daily to sell produce, but the market is not frequented by outside traders. A number of people indicate that there is little point trying to sell any produce locally because everyone is producing the same crops. Individuals with crops to sell have to transport them to Awinare and Kwabeng.

Boma is not a center of development activities. It has no active groups interested in agricultural activities, and the once-powerful Ghana National Farmers' Council does not have a functional presence. The local crop extension officer based in Awinare reported that he had tried to encourage Boma people to form a cooperative that would give them access to credit and other agricultural services, but they were not interested. In contrast, he describes the farmers of Awinare and Kwabeng as being more responsive to his recommendations. The most active civic groups appeared to be those that were church-related.

Although Boma gives the appearance of being isolated, links with formal research and extension activities are strong. The village lies within 50 kilometers of the main cocoa research station at Tafo and the extension training center at Bunso. Palm oil research and extension is based in Kade, which lies farther east along the same laterite road that enters Boma. A number of our respondents had visited these stations and their substations. Although the crop extension officer was somewhat disappointed with farmers in Boma, he provides individual assistance on request, for example, in lining and pegging for palm planting or with garden egg production. At the same time, a number of our respondents were working or had worked for the Ministry of Agriculture. Because cocoa extension is still active, when we started visiting farms, people immediately assumed we were inspecting farms for swollen shoot damage and offered their farms for inspection in the hope of receiving compensation or some other form of "encouragement."

The second major site within Eastern Region, Ghana, is the village of Suminakese, situated 91 kilometers from Tafo, northeast of Nkawkaw (a major commercial center on the main road linking Accra with Kumasi). Although the village is located about 30 kilometers down a winding, rolling, and gradually climbing laterite road, because this is an important food-producing area, and significantly closer to Accra than a number of other tomato-producing areas, vehicles ply the road regularly throughout the day.

Suminakese has a population of about 5,000 and is apparently still growing. Most residents are Kwahus who moved to these farm sites to produce cocoa at the turn of the century. Adjacent to Suminakese is the hamlet of Brafoa, which, although said to have been settled earlier is now a sleepy, small village with around twenty compounds, some of which appear to be unoccupied. Some residents of Brafoa have gone to live in Suminakese, and the Brafoa village primary school is now closed. Suminakese is said to have spread closer to Brafoa over time, the first settlement having been close to the river in the valley bottom. In contrast to Brafoa (and Boma), Suminakese appears to be a lively, thriving community with many small bars and cooked-food establishments, small shops, and a well-attended, twice-weekly market. At the same time, it is important to note that the majority of Kwahus living in these sites continue to regard themselves as immigrants, and although they reside here and send their children to school in the village, festivals are celebrated and houses are built in their "hometowns."

Farming is the main occupation in the area. Suminakese continues to have tracts of forestland where cocoa, oil palms, plantain, and root crops are grown. People from Suminakese also rent land in Brafoa, which is largely guinea savanna, for vegetable and groundnut production. Although the vegetable crops are mostly sold on market days and the road is poorly maintained and dangerous, traders arrive in taxis and trucks to make purchases throughout the harvesting season.

Suminakese is historically known for its commercial tomato production, and an old sign for a farm cooperative marks the turnoff to the village from the main artery road from Nkawkaw. One of the primary subjects raised in our interviews was the problem of tomato marketing, in particular the low prices received by producers. In contrast, widespread interest was expressed in onion production, particularly by women. There is no resident agricultural extension staff, and a number of people in Suminakese report going directly to the ministry in Nkawkaw for information.

Unlike Boma, Suminakese appears to have an active Local Development Committee that serves as the main link with formal development organizations and organizes communal work. World Vision, an NGO, operates borehole and school construction programs in the area, while the Ministry of Health runs a child health program. People are encouraged to bring their children for weighing and screening through the provision of food aid (dried milk and flour). As elsewhere in Eastern Region, church groups are active in Suminakese.

Fieldwork Sites 5 and 6: Brong Ahafo Region,
Ghana: Pamdu Village, Nkoranza Traditional Area,
Kintampo District; Awisa Village, Wenchi District

The south-central part of Brong Ahafo Region in central Ghana marks
the boundary between the forest to the south and the savanna to the
north. As such, it contains a wide range of vegetation types, from resid-
ual pockets of high forest to gallery forest and grass savanna. Rainfall
averages approximately 1,200 millimeters per year over nine months.
Brong Ahafo Region is one of the least densely populated regions of
Ghana: Amanor (1993b:15) reported the population density of the region to be nineteen and thirty-one persons per square kilometer for
1970 and 1984, respectively, while the Ministry of Agriculture (1991:6)
cited a figure of thirty-seven persons per square kilometer for 1990.
The total rural population of the region was estimated at 1.06 million
in 1990.

Brong Ahafo is a major food-production area for Ghana. The southern part of the region marks the northern fringe of the cocoa belt. However, many old cocoa plantations were destroyed by severe and widespread bushfires in 1982–1983, which resulted in the availability of large
areas of land that have subsequently been put into maize production.
Since the 1950s, there has been significant public-sector investment in
mechanized farming in Brong Ahafo, and while the mechanization
schemes themselves failed, present-day farmers have made use of the investment in infrastructure to expand and intensify production (Amanor
1993b:13). In addition to maize, major crops include yam, cassava, rice,
and groundnuts, while a whole range of minor crops are grown including
vegetables such as tomatoes and onions.

The old cocoa economy of southern Brong Ahafo was based on the
influx of migrant labor from northern Ghana. Current food-production
activities are also heavily dependent on labor from the north, and there
is active migration of Dagati-speaking farmers (Wala, Nandom, and
Lobi peoples) into this area. Many migrants quickly gain access to land
to farm in their own right, and most villages have a special area, or
zongo, where the migrants reside. The majority of the population of
south-central Brong Ahafo is Akan-speaking Brong.

Both Pamdu and Awisa villages represent comparatively well-endowed, dynamic agricultural environments where there is active interest
and investment in commercial food production, in particular, vegetable
production. Agriculture appears to offer a viable and attractive alternative to off-farm income-generating opportunities. Although both sites
are well served by transportation infrastructure, neither appears to be a
major focus of agricultural extension or research activities.

The village of Pamdu is located on an excellent tarmac road 32 kilometers north of Techiman (which is considered to be the largest wholesale food market in Ghana) and 27 kilometers south of Kintampo. The village consists of approximately 140 residential buildings, perhaps 40 of which are inhabited by northerners and the remainder by Akan. The total population is between 2,000 and 2,500 persons. Pamdu has two schools, two bars, two small kiosks, a clinic, a small chemist shop, and two privately owned grinding mills. At the north edge of the village is the settlement of Bosom Krom, which consists of approximately eighty households of northern migrants and ten Akan households. Although officially tied to the neighboring village of Paninamisa, Pamdu clearly has a very strong relationship with Bosom Krom. The residents of "Old Pamdu" migrated to the present site between 1968 and 1979 following the construction of the Kintampo-to-Techiman road.

Pamdu has good road links with major administrative and market locations. A wide variety of vehicles come to Pamdu on a daily basis to pick up tomatoes, maize, and other produce. At the present time, there is no extension agent based in the village, nor does it appear that an agent makes regular visits. Nevertheless, there is an individual who lives and farms in the village and who worked as an agricultural extension agent for a number of years, including a period based in Pamdu between 1989 and 1991 with the Global 2000 program. While some seeds and agrochemical products are available in Pamdu from private individuals and at the kiosks, it is most common for farmers to go to Techiman or Kintampo to purchase seeds, sprays, and fertilizer.

The major crops grown in the area are maize, yam, vegetables, cassava, and tobacco. Pamdu used to be noted for the production of tobacco, and there are still tobacco barns within the village and to the north. According to one informant, around 1971 the people of Pamdu started planting tomatoes in the tobacco fields after harvest. At that time, they did not use fertilizer or chemicals on the tomatoes. Subsequently, they experienced certain problems producing good quality tobacco, and so many young men turned to tomato and vegetable production in earnest. Several informants indicated that they went into tomato production after finishing school or an apprenticeship because it was clear that they would be able to make money. Within the village, there are four tractors, several maize shelling machines, several small irrigation pumps, and many hand pump sprayers for chemicals.

Tomatoes can be planted at four times during the year, in March, July, September, and January. Only farmers with access to land near the river can plant tomatoes in January as irrigation is essential. Farmers refer to a number of tomato varieties, including "Paul," "Rhino 70," "Quick Money," "Yokahama," "Power Dwarf," "Power," "Power Rhino,"

"Power Rasta," "Power—the short one," "Power—the tall one," "Co-coa," "Cocoa Power," and "Akua." Discussion as to the characteristics of these varieties is detailed and animated.

Tomatoes are shipped in wooden crates that belong to the traders and that are distributed in advance by them. Getting access to the crates is a major step in successfully selling the crop. The price of tomatoes fluc-tuates significantly: At the time of our fieldwork, farmers were receiving per crate just half of what they were only one month before.

There are a number of both active and dormant groups and asso-ciations whose activities relate more or less directly to agriculture. The Pamdu Tomato Growers Association (PTGA) and the Pamdu Vegetable Producers Association (PVPA) were formed around 1986 in order to attract traders to the village. Both groups appear to have been based largely on personal and family links, and neither group is now particularly active. Other groups have been formed to market crops, such as beans, maize, and yams, and to access credit from for-mal sources. Some of these groups receive support from formal de-velopment organizations.

Many Pamdu residents participated in the Global 2000 program which focused primarily on the production of maize. Apparently the re-payment of credit was poor, and the program was consequently with-drawn. Technoserve, an international NGO, is presently active in Pamdu promoting the storage and marketing of maize through farmer groups.

Research at the next site, Awisa village, was through a collaborative arrangement with the WFSTP, and the description that follows draws heavily on Waldie (1995). Awisa village is situated 5 kilometers north of Wenchi town and straddles the road between Wenchi and Wa. The 1993 National Electoral Commission's Register of Voters indicates that there are 721 adults aged eighteen years or above in Awisa town (not includ-ing the *zongo* situated on the northern outskirts of town, which includes about forty-five houses, or the outlying hamlets). The town has been growing rapidly since 1985.

In addition to its proximity to Wenchi, Awisa has good access to the major towns of Techiman and Kumasi. Its cosmopolitan feel is height-ened by the fact that a number of people from Wenchi town farm in the area and by the daily presence, from November to March, of tomato traders from Kumasi and Accra.

From an agro-ecological perspective, Awisa is similar to Pamdu and can be described as derived savanna. Land use patterns have changed a great deal within living memory. The major bushfires of the early 1980s are reported to have cleared many hectares of unproductive cocoa. The major crops are now maize, yam, cassava, groundnuts, and vegetables. Some tobacco is also produced, but few farmers speak enthusiastically about tobacco production as the work is difficult and the farmers feel dependent on the tobacco company for inputs and marketing.

Migrants from the north have a clear sense of their farming being different from that of the local residents. They report having met cassava and bambara groundnut here, but they brought skills in guinea corn, cowpea, and groundnut farming from their home areas. There does not appear to be any shortage of available land for farming, and access to land is frequently gained in exchange for work or through renting. Although the cost of renting land may prove a constraint to some farmers—in particular, the Dagati, who have no customary claims to farm in Awisa—the general view expressed by inhabitants is that land is available. There are two tractors, three power grain mills, and one irrigation pump in the town.

Tomato-production activities are broadly similar to those in Pamdu except that there is little planting in January as the supply of water for irrigation is limited. The period of peak tomato production is between December and March. Due to the general ineffectiveness of their formal associations, farmers often organize themselves into groups and pay a contribution to send someone to entice a trader to come from Kumasi or Accra. It is said that special relationships develop between farmers and particular traders and that, for example, a farmer might receive a commission or an advance payment for tomato production (this was also true in Pamdu). There is, however, some skepticism about these kinds of relationships. The lack of confidence, coupled with the anxiety to "find the market," frequently leads to disputes, not only between traders and farmers, but among the farmers themselves as well.

Awisa does not have a resident agricultural extension officer, but this role appears to have been taken on by one resident who works for the Crops Research Institute. This individual also sells agricultural chemicals. One or two other residents have received formal agriculture training. The Subinsu Agriculture Project, situated another 10 kilometers along the road to Wa, appears to have influenced the farmers of the area through provision of agriculture advice and of maize and cowpea seed on credit. The WFI, although much closer, was barely known. The Pioneer Tobacco Company is another source of formal agricultural information.

There are four functioning associations in Awisa: the Rural Youth Association, the Tomato Growers Association, various cassava groups, and the Dagati Committee. The Awisa Rural Youth Association grew out of a Food and Agriculture Organization (FAO) initiative in 1991, and is linked to an association in Germany. A tractor has been provided, which is hired out to association members, and individual loans are made to members from the rental proceeds. There are 180 members in

Wenchi District, 45 of them from Awisa (20 of whom are women). The Tomato Growers Association was also started in 1991, but at the time of our field research, it was basically inactive. There are three or four cassava groups, each with ten members. The groups have nothing directly to do with cassava as a crop, but are simply groupings of farmers, organized to gain access to credit, supposedly for investment in agriculture. All that remains of the Awisa Food and Vegetables Co-operative Society is a signboard. Global 2000 also operated in the area around 1989 when an extension officer attempted to organize farmer groups, but all these have since collapsed.

Site 7: East Anglia, United Kingdom

The region of East Anglia, comprising the counties of Norfolk and Suffolk with parts of Cambridgeshire and Essex, is located in southeastern England to the north and east of London. This is one of the most intensively farmed regions of the United Kingdom, and its agriculture, which is large-scale and highly mechanized, is dominated by arable crop production. On average, the area receives 600 millimeters of rainfall per year, this being distributed more or less evenly over the year, and the growing season averages 240 days. The area is characterized by a complex series of soil sequences, ranging from Marine Clays and Fen silts to Sandy Boulder Clays and Chalk and Limestone Soils.

The major cereal crops are wheat and barley (covering 50 percent of the arable area); other important crops are sugar beets, potatoes, oilseed rape, and peas. Although the region is not considered a livestock area, there are 150,000 cattle, 130,000 sheep, and 1.2 million pigs in Norfork and Suffolk (Murphy 1995:222). In Norfolk and Suffolk there are approximately 9,300 agricultural holdings: 30 percent of these are between 40 and 200 hectares in size and 11 percent are over 200 hectares (ibid.:221). There are essentially three types of farmers in East Anglia: owner-operators (with or without additional management), tenants, and contract managers.

Since the mid-1980s, farmers in East Anglia, like their colleagues throughout Europe, have had to contend with major changes in international markets and trade, the ongoing reform of the European Community's Common Agricultural Policy, and the public's increasing concern with issues such as environmental pollution from agriculture, food safety, and animal welfare. Perhaps one of the most important of these changes is the requirement that 15 percent of cultivated land must be "set aside." Reductions in cultivated acreage, limitations on particularly profitable crops such as sugar beet, more stringent regulations concerning water pollution from agriculture, and increasing competition from eastern Europe mean that farmers in East Anglia must become increasingly efficient in the use of resources.

Farmers gain access to information through many channels, including their neighbors, radio, the press, private consulting agronomists, and company sales representatives. Since the mid-1980s, the government-supported extension service has been largely privatized, and most services are now on a fee basis. There are a number of research institutions that directly or indirectly contribute to the development and testing of technology for farmers in East Anglia. While some of these research activities and institutions are funded directly by the government, others are funded through levies on particular commodities (e.g., cereals and sugar beet), and still others, such as crop variety and agrochemical development, are funded primarily by private industry. One unique element of the agricultural research scene in East Anglia is the Morley Research Centre, which has remained independent and farmer-funded since its establishment in 1908 (Hutchinson and Owers 1980; McClean 1991). Morley's farmer-members play an important part in setting its research direction and agenda, and they receive the results of research through a variety of channels including limited-circulation bulletins and field days.

6

Some Characteristics of Farmers' Experiments

The results of our field research are presented and discussed in this chapter and Chapter 7. This is done first by looking at the topic or subject of the experiments identified by farmers and relating these to the different contexts within which they were carried out. In presenting the results in this way, we do not claim that the experimental topics within each of the sites reflect, with any degree of precision, the characteristics of an underlying "population of experiments." Nevertheless, the examples provide some indication of the concerns of farmers and some insight into the variability and scope of farmers' experiments under different conditions.

Subsequently, as far as the data allow, the experimental processes and methods are described, and we attempt to link the characteristics of the experiments with certain socioeconomic characteristics of the farmers who reported them. Here, our objective is not to provide a definitive description of the personal, social, cultural, ecological, or epistemological bases of farmers' experiments; rather, it is to develop an analysis of the nature and characteristics of farmers' experimentation as a phenomenon with which formal agricultural research might *regularly* interact. The results of such an analysis will be essential for our evaluation of the synergy hypothesis.

CHARACTERISTICS OF FARMERS' EXPERIMENTS IN AFRICA

Over all sites in Africa, 155 examples of farmers' experiments were cataloged. Although the examples vary quite considerably, by definition they all share one common element: Something "new" or different was tried by a farmer who was subsequently able to recount some details of both the process and the outcome of the experience. For purposes of illustration, four of these examples of farmers' experiments are presented in

Table 6.1. It is important to note that the examples of farmers' experiments that we cataloged are similar to farmers' experiments reported by Millar (1994); Budelman, Buchukundi, and Mizambwa (1996); Stolzenbach (1994); and others.

Table 6.1. Examples of Farmers' Experiments from Africa

Site/Interview	Example
Ghana/g159	In July 1994, she planted about one-eighth of an acre of watermelon for the first time. She had seen watermelon growing on Kojo Yeboah's farm, but he was not there at the time. She collected seeds from people she saw eating watermelon. She does not know about fertilizer application and guessed about planting distance and date. Her crop did not do well, and as a result, she sought advice from a friend whose father also grows watermelon. The friend advised her to apply fertilizer two weeks after planting, but it was already too late. Although her first attempt was a failure, she intends to try again and will also try to extend the area. She has not visited any other farms to see watermelon being grown.
Ghana/g222	"This is an experiment farm." He treated a couple of rows with a liquid starter fertilizer that he made himself from granular fertilizer. He applied the liquid three days after transplanting; the adjacent rows received regular granular starter only three days before the interview. There is a dramatic difference in the growth and color. He says the idea came from his own head—he will pluck the first flowers from the more vigorous plants so all the fruits mature at the same time.
Zimbabwe/z4	Last year, she made ridges with a hoe and planted sorghum on them. It was her first time using ridges, and the first time planting sorghum with a hoe: She normally planted following the plow. The crop was healthy and the harvest good. The only problem was that the field was too small! This year, because of the poor early rains, she plans to extend the same ridges.
Kenya/k11	He used to plant bananas by digging only a small hole, but he noted that the plants did not do very well. He remembered that during earlier periods of ethnic conflict, people dug large holes in which to hide. Later, when these holes eventually filled with leaves and garbage, some people planted bananas in them and they did very well. So he dug a large (4 feet in diameter) hole and filled it with loam soil and manure, then planted bananas. They are doing well.

Note: Here and throughout this book, we refer to the interviewees/respondents in our fieldwork by the numbering system shown above (i.e., "g159," "g222," "z4," and "k11").

Topics of Experimentation

Each of the 155 examples we cataloged was categorized as to its topic or subject matter, and the frequency distribution of the topics is shown in Figure 6.1. The examples of experimentation address a wide range of questions and issues, from the purely agronomic, such as crop variety selection and planting density, to the social, economic, and institutional, including labor and marketing arrangements. The most common agronomic topics are land preparation; new varieties; new crops; plant spacing, density, and fertilizer; and soil fertility. Together, these account for 75 percent of all examples; the nonagronomic topics account for only 5 percent of the examples.

The topics addressed by the farmers differed among the research sites. Thus, the examples from Chivi, Zimbabwe, are dominated by land preparation (49 percent), while those from Brong Ahafo are dominated by new crop, new variety, and plant spacing and density (65 percent), and those from Eastern Region by new variety, plant spacing and density,

Figure 6.1
Frequency Distribution of Topics, All Sites (*N* = 155)

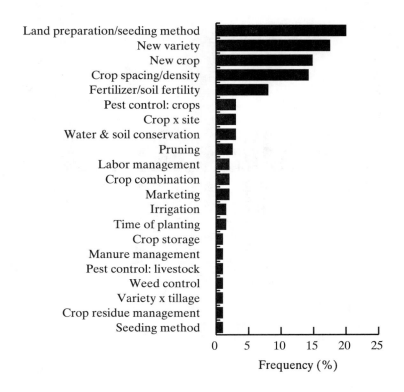

90

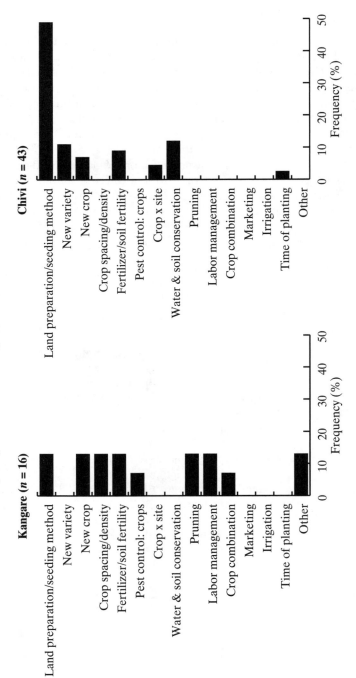

Figure 6.2
Frequency Distribution of Topics, Individual Sites (*N* = 155)

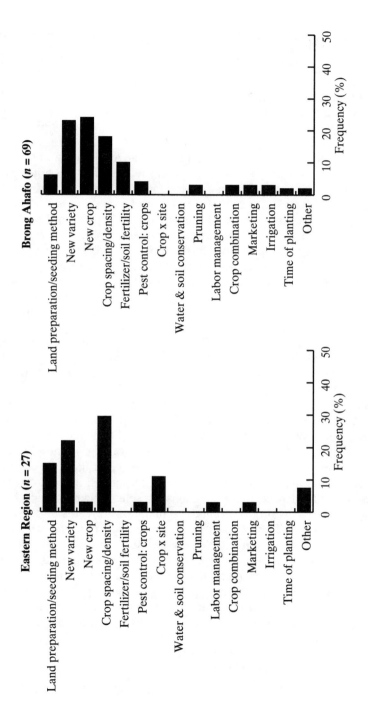

and land preparation (67 percent) (Figure 6.2). Only in Kangare is there no dominant subset of topics.

At first glance, these data certainly appear to confirm the proposition that farmers do engage in activities that, within the bounds of our definition, can be described as experiments. But what do we learn from these findings, particularly in relation to the topics of the experiments? How much significance should be placed on the frequency distributions of topics associated with the examples?

A number of factors clearly influenced the distribution of topics, and there is little question that the field methodology itself was among these. Through our "direct approach," we sought examples of experiments carried out at some time in the past and hoped to provide an opportunity for informants to move beyond the immediate bounds and concerns of the cropping calendar. For example, an elderly woman in Boma (g35) told how years ago she and her husband were the first to try to plant on the hills near the village. Although everyone said cocoa would not do well there, they planted some and it did well. Now everyone plants there.

In contrast, our "indirect approach" focused on the difference between "normal" and "actual" practice for identifying experiments that were ongoing at the time of the interview, usually at the level of an individual field or plot. Examples elicited through the indirect approach necessarily reflect the particular point in the agricultural calendar at which the interviews took place, and thus the timing of the fieldwork, to a large extent, circumscribed the topics of the examples. In Chivi, for example, interviews were conducted at the beginning of the cropping season when the informants were actively engaged in land preparation and the planting of millet, sorghum, maize, and groundnuts. The fact that nearly half of the examples drawn from that site concentrated on land preparation and seeding method should not, therefore, be particularly surprising.

The Chivi area is subject to recurrent drought; indeed, during the time of our fieldwork, there was considerable concern among farmers about the late arrival of the rains, and many of the fields that we visited had already been planted a second time due to poor germination or desiccation of cereal seedlings following germination. Farmers told of experiments with land preparation, but also with soil and water conservation, and a number of farmers highlighted the fact that the drought was forcing them to try new things. The drought of 1992–1993 resulted in the loss of a considerable number of livestock, including draft animals, and many farmers have thus been forced to reduce their level of mechanization. Some farmers who were interviewed, for example, claimed that they were farming manually (i.e., with a hoe) for the first time in twenty years. This reduced mechanization has also been a stimulus for experimentation and adaptation: During our field research, we observed people preparing land for planting using hand hoes, draft teams of two donkeys, two donkeys plus two cattle, two cattle, four cattle, two heifers plus two cows, tractors, and a moldboard plow being pulled by the people themselves.

But there are clearly some limits on the range of possibilities and the ability of particular individuals to actively experiment in the face of such difficulties: While one farmer reported that "when the situation is bad enough one will try almost anything" (z71), a neighbor declared, "This year I cannot play around with a trial" (z12).

A third variable was particularly influential in determining the subject of experimentation in the Chivi site. It was observed earlier that ITDG, our collaborating institution in Zimbabwe, was working directly with the selected communities in Chivi. In Ward 21, ITDG is actively promoting the idea that farmers should test a range of soil and water conservation techniques, including tied-ridges.

In Ghana, with its greater rainfall and a longer growing season, the agricultural calendar is considerably more flexible than in Chivi. At the time of the interviews in early January 1995, farmers at the two Brong Ahafo sites were engaged in a wide variety of tasks, including harvesting maize, yam, and tomatoes; preparing land; nursing and transplanting tomato seedlings; and spraying. In addition to rainfed crops, farmers in Brong Ahafo sites also produce vegetables under irrigation during the dry season. This range of activity creates considerable room for experimentation, which is evident in the examples from these sites.

The predominance of examples that focused on new varieties, plant spacing, and density at the Ghana sites probably reflects the very competitive, commercial nature of food crop production, particularly in Awisa, Pamdu, and Suminakese. The singular importance of the tomato crop in Awisa and Pamdu, coupled with the fact that tomatoes can be produced during four distinct seasons, places the producers in a relatively weak position vis-à-vis the traders. Producers here see some considerable marketing advantage in growing fields with a mix of fruit types. It was reported in Pamdu that in cases where boxes of tomatoes offered for sale are predominately of one fruit type, traders may use this fact to drive down the price with claims to the effect that "this type is not in high demand in the market." While farmers claim to recognize as many as thirteen varieties of tomatoes, the agronomic and commercial characteristics (e.g., vigor of vegetative growth, firmness of fruit) of even the most commonly grown varieties are hotly debated. Seed of few if any of these varieties is ever produced or marketed through commercial or government channels, nor are "pure" seed stocks maintained by local farmers. Thus, the only way to assure the availability of a range of varieties and types is through testing and experimentation with small seed lots.

Tomato varieties were also the subject of much discussion in Suminakese. It is interesting to note that one farmer in Suminakese (g115) had come to the conclusion that it was precisely because of the variability of his village's tomatoes that he and his neighbors were suffering at the hands of the traders. This individual provided an example of an ongoing experiment in which land is being sought for group production of tomatoes with strict quality controls. In general, farmers reported a loss of marketing advantage with tomatoes and expressed greater interest in onions and garden eggs. In contrast to tomatoes, no varietal differences were reported for these two crops, and no experiments were reported.

It is clear that the constant movement of farmers and traders in and out of the village facilitates the introduction of small quantities of seed of new varieties and new crops. Several informants recalled trying seed that had been introduced in this manner, including a woman from Pamdu (g225) whose husband returned from a trip to Accra with a tin of the variety "Paul," which he had purchased at the Department of Agriculture. In Boma and Suminakese, a number of individuals reported collecting cassava cuttings and suckers of plantain from varieties that they had seen and liked, but these individuals reported these events as simply collection activities rather than tests of something new to be observed and recorded for future action.

Concerning tomatoes, key individuals were reported to have played a central role in the introduction of varieties that had transformed the production system. In Pamdu, the tomato variety "Power," which seems to be the most widely grown variety and which many consider to be the key innovation upon which the current dynamic tomato economy is based, is said to have been introduced into the village by a member of a prominent family who had a university education and had traveled internationally. One individual in Suminakese, a former driver who had settled in the village within the previous ten years, was also widely reported to have transformed tomato production through the introduction of new varieties.

Once varieties are introduced, experimentation begins in earnest, as is evident in several of the reported examples. One informant in Pamdu recalled the first time he tried the "Power" variety in 1985, and the fact that he subsequently experimented with a modification to the planting density that had been recommended:

> Ofosu first brought the variety "Power" to Pamdu in 1985 with a recommended intrarow spacing of "one plant per cutlass length." After planting as recommended for one year, our informant increased the density to three plants per cutlass length, which was how most people had planted what was to that point the most commonly grown variety. During this trial year, other farmers came to his farm and said that with

such a high density there would be too much foliage; but in the event, he harvested a lot of fruit, and now he claims that they all plant "Power" at this density. (N.B.: In fact, even among those farmers who claim to plant "Power," intrarow spacing varies considerably, apparently depending largely on soil fertility status, and is still an important subject for experimentation.) (g259)

The lack of any dominant subject for experimentation in Kangare is not easy to explain. Certainly, there is no crop in Kangare for which production and marketing is particularly competitive, and it was also our impression that there was no single, widely acknowledged problem around which experimentation would naturally, if perhaps only temporally, galvanize. The keen interest in dairying at this site has already been noted, and one individual did report experimenting around this activity.

In this light, it is interesting to note that it was in Kangare that we were told repeatedly, and apparently in earnest, that there was no real reason to try new things or seek advice because, "we have always grown maize, there is nothing we do not know." Very similar discussions took place in Boma, Ghana.

In fact, our framework that sought to contrast "normal" with "actual" practice was most problematic in Kangare and Boma. Weeding was the major ongoing agricultural activity at the time of the interviews in these sites, but no one from either site reported experimenting with weeding. In addition, in Kangare, some interviews were conducted in maize fields that contained plants at *all* stages of growth, from immediately postgermination to nearing maturity. In such a situation, it is almost impossible to talk of something being "normal" at the level of an individual activity such as planting or weeding. Perhaps the more important element of normality in this situation is the continuous planting itself, as opposed to the way the planting is done on any particular day. Similar circumstances were encountered in Boma in crop fields where food was being grown for home consumption. Here, not only was there considerable variation within an individual crop, but also each field contained numerous different crops, all at various stages of development. These fields were invariably farmed by women.

The topics of experiments are also determined by the very nature of crop production itself. In most cases, crop production consists of a more or less predetermined sequence of stages and decisions, starting perhaps with land preparation and continuing through crop and variety selection, planting at a particular spacing and density, weeding, harvesting, storage, and marketing. This sequence, and the rather limited number of major decision points that it affords, particularly when the process is situated within a given agro-ecological and economic context and an existing production system, effectively defines the potential domains for experimentation. It is also true that when isolated, most of these decision points have high "trialability": It is relatively straightforward to modify the direction or depth of plowing, substitute one variety for another, or increase or decrease plant density. On the other hand, experiments involving more fundamental, systemwide change, or institutions and social relationships, might not be expected to arise out of the normal sequence of farming, may have much lower trialability, and will probably entail more risk because of their scale and/or complexity. For example, the experiment referred to earlier where an individual in Suminakese attempted to organize group-based production of tomatoes is both socially and technically complex. And the following example of a woman in Brong Ahafo who tried to establish a new marketing arrangement was complicated by the fact that she needed a partner to spread the risk; according to our informant, it was her partner's actions that were responsible for the experiment's disastrous outcome:

> Akua purchased fifty of her own wooden tomato boxes, and her plan was to buy tomatoes in Pamdu and rent a lorry [truck] to transport them to markets in the south. Because she could not carry all the risk herself, she entered into an agreement with another woman from Pamdu. In the event, her partner cheated her, and she not only lost a lot of money but is still in debt to the lorry driver. She says that now she is scared of the tomato business because it is too risky: "You can lose everything." She abandoned the tomato boxes in a nearby village: "If someone has stolen them I do not know." (g224)

Motivation and Methods

Overall, and at each of the sites in Africa, the vast majority of examples of experimentation (65–95 percent) appeared to be proactive rather than reactive (Table 6.2), which is to say that most examples indicate a degree of planning and willful action as opposed to improvisation or unplanned reaction to hazard. The predominance of examples with a proactive character gives added credence to the notion of small-scale farmers as active experimenters who purposefully set out to solve particular problems through the testing of various options.

A strong bias toward examples of proactive experimentation might have been expected if the identification of examples relied solely on the direct approach (i.e., through a direct discussion of "trying new things" or in response to a straightforward request for examples). However, the indirect approach, with its focus on ongoing activities and the use of the distinction between "normal" and "actual" practice to help pinpoint experiments, should be less dependent on a respondent's memory of having actively taken steps to try or test something new. It is also clear that experiments in the most common subject areas—including land prepara-

Table 6.2. Characteristics of Farmers' Experiments (% within sites)

Characteristic	Kangare	Ward 21	Eastern Region	Brong Ahafo	Overall
Motivation					
Proactive	73	81	75	94	85
Reactive	27	19	25	6	15
n =	15	42	24	68	149
Method					
Without control	69	58	84	52	61
With control	31	42	16	48	39
n =	13	38	19	50	120
Source					
Trying what has been observed or suggested	43	35	35	65	51
Trying something that has been actively promoted	0	38	0	12	15
Own idea	57	28	65	23	34
n =	14	29	17	60	120
Outcome					
Something novel	31	12	12	3	10
Rediscovery; minor modification	63	81	89	88	83
Major modification	6	7	0	9	7
n =	16	43	26	69	154
Topic					
New crop	13	7	4	25	15
New variety	0	12	22	23	18
Spacing/density	13	0	30	17	14
Fertilizer/soil fertility	13	10	0	10	8
Land preparation/seeding method	13	50	15	6	20
Other topics	50	21	30	19	25
n =	16	42	27	69	154

tion, new crop, and new variety—by their very nature actually demand some degree of forward planning.

The predominance of proactive experiments means that most of the examples of farmers' experiments that we identified were neither simply serendipitous nor so deeply embedded in local culture or the practice of farming (i.e., *art de la localité*) as to be indistinguishable, in the mind of the farmer, from everyday farming. Rather, they were planned and purposeful tests. Stolzenbach (1994:155) used Schön's (1983) notion of "reflection-in-action" and the idea of agriculture as "a performance" (Hanks 1972:28; Richards 1989) to suggest that it may be "difficult to talk of an 'experiment' as a special action." These data do not lead us to the same conclusion. Instead, while trying things may well be a "natural" and integral part of farming, many farmers appear to go about their experiments in a conscious and planned fashion. Indeed, it is difficult to see how they could do otherwise and still gain the knowledge they need to survive in dynamic and often difficult environments. The common observation that farmers often give local names to crop varieties introduced from the formal research system and actively adopt elements of technological packages in a "pick-and-mix" fashion can be accounted for only by a process of proactive, purposeful experimentation, which we suggest is also clearly demonstrated by our findings.

The idea that the "indigenous" knowledge of small-scale farmers, and of rural people more generally, is orientated, generated, and structured in ways that are somehow both different from, and largely incompatible with, "formal" knowledge is often an important element of perspectives rooted in populist political and cultural anthropological traditions. As illustrated by these examples of farmers' experiments, however, such a view does not reflect the fact that in the latter half of the twentieth century, many farmers in Africa either use or know of techniques and crop varieties developed and promoted by the formal research system, as well as other products such as fertilizer and pesticides. In addition, many farmers have some formal education and have been exposed to a variety of extension programs and development projects, with their accompanying training, on-farm trials, and demonstration plots.

In terms of the methods used in carrying out the experiments, over all sites, 61 percent of the examples of experimentation for which data are available appear to have no obvious provision for a control or direct comparison, while 39 percent include a direct, side-by-side comparison (Table 6.2). There is some variation in this pattern over the four sites, with Brong Ahafo and Zimbabwe having a slightly higher, and Kangare and Eastern Region a slightly lower, proportion of examples with an obvious control. These comparisons usually entail a small "experimental plot" within a field that is farmed in the usual manner, or the field being split with the "experimental treatment" applied only to one part.

However, these data give a somewhat distorted view of this characteristic of farmers' experimentation. In the first place, because of a lack of specific information to the contrary, a number of the examples that involve trying a new variety were classified as having no obvious control. It seems likely, however, that in most cases new varieties would be grown near other plots of the same crop that would act as a control. Second, it is not obvious how the testing of new crops, which accounts for 15 percent of all examples, could easily include provision for a control or comparison. When the new crop examples are eliminated, examples with an obvious control increase to 45 percent of the total. Third, and perhaps most significant, it is clear that some farmers use, and have considerable confidence in their sense of, an internalized "historical control" that is based on their accumulated understanding of the past performance of a particular field or crop and the major factors affecting that performance such as rainfall. This is clearly illustrated by the following example from Zimbabwe:

> Francis made a trial because he wanted to know which crop was best suited to a particular soil type on his farm. The first year, he planted millet in the field without the application of any manure or fertilizer, and he harvested five bags. In the second year the same plot yielded twenty-five bags of maize. During the interview, it was indicated to him that his trial confounds the effects of crops and years, but he replied that as far as he was concerned, the years, and specifically the amount and timing of the rainfall, were similar enough to be compared. He subsequently tried the same type of experiment with sunflower and sorghum, but on a different plot. He is now working with another two plots that he planted last year with groundnuts and maize. He should rotate the crops this year, but he is afraid that the peculiar rainfall situation may make the comparison difficult. (z54)

However, even without taking account of this use of a historical control, approximately half of the examples for which data are available show some provision for controlled comparison. As the majority of the examples seem to be proactive experiments consisting of relatively minor modifications to existing production activities, it seems both plausible and logical that they should incorporate "new vs. old," "with vs. without," or "before vs. after" comparisons. The example cited above also illustrates the fact that some farmers actually quantify the treatment effect by measuring the yield of different plots. Another farmer in Zimbabwe reported that he learned how to compare yields from plots of the same size through his association with demonstration plots managed by the extension service.

These data lend little support to those who have suggested that farmers' experiments are by their very nature haphazard or usually entail an "all-at-once" change and thus provide no basis for comparison or evaluation (Gubbels 1988:12; Millar 1993:48; Connell 1991:218). First, it is important to distinguish between an all-at-once change to a previously untried variety or technique and an alternating use of several new or previously tried techniques. Second, the obvious implication of most experiments being small in scale, focused on relatively small changes in a limited number of existing agronomic practices, and to at least some degree planned is that, in many cases, a meaningful control or comparison is likely to be close at hand. "Haphazard," then, is not an appropriate descriptor of the farmers' experiments we documented.

When seen as individual acts, these unreplicated, often confounded tests, most of which rely on unquantified comparisons, are not particularly powerful or efficient for determining the effects or the value of alternative treatments. Nevertheless, as these tests are often repeated over several years, at least some farmers are able to perform a kind of intuitive covariance analysis (i.e., adjusting and interpreting the results in the light of his or her experience of a particular crop or field over different years). In this light it seems reasonable to accept that these tests yield valuable, site-specific information with minimum risk and cost. They essentially provide farmers with the confidence, born through hands-on experience within a known context, to move the production system in response to a dynamic array of constraints and opportunities.

Sources of Ideas and Techniques and Outcomes of Experiments

The most common sources of the ideas associated with the examples of farmers' experiments were "trying what was observed or suggested" (51 percent) and one's "own idea" (34 percent) (Table 6.2). "Trying something that has been or is being actively promoted" accounted for only 15 percent of the examples. The importance of the different sources varied greatly between sites: In Kangare and Eastern Region, approximately 60 percent of the examples were reported to be based on the experimenters' own ideas, while in Brong Ahafo and Zimbabwe, the farmers' own ideas were associated with less than 15 percent of the examples. In Zimbabwe, 38 percent of examples were reportedly based on trying something actively promoted, while this category accounted for less than 11 percent of examples at the other sites.

We can begin to understand some of these differences by referring to the context within which the experimentation is taking place. Farmers at the Kangare and Boma sites are generally less directly dependent on farming, crop cultivation is less commercialized and competitive, and there are few widely acknowledged problems when compared to other sites in Ghana and to Chivi, Zimbabwe. In these circumstances, we would argue, much of the drive to experiment is internal rather than external to the individual, with farmers' experiments essentially reflecting individual interest in trying ideas that come to mind. Further, the individual's interest in and need to experiment may not be well supported by the larger farming community. The experiments in Kangare and Boma may thus represent the "curiosity" experiments that Rhoades and Bebbington (1995) suggest are characteristic of stable environments. In contrast, at the sites dominated by highly commercial and competitive production, and in Chivi with the continuing problems of rainfall and food insecurity, there is much greater motivation to seek new ideas from one's neighbors and from other external sources including the formal extension services.

Although there was some level of formal extension activity at all sites, Chivi, where our research took place within the ITDG project area, was the only site where a significant proportion of examples were classified as originating from "trying something that has been or is being actively promoted." Particularly in relation to experiments concerning land preparation and soil and water conservation, ITDG was repeatedly cited by respondents as the origin of the idea being tested. It is interesting to note that ITDG's activities in Ward 21, despite its focus on farmers' experimentation, were often referred to during interviews as "lessons," which is the same word used to describe more traditional extension activities undertaken by state extension services. This raises the question of whether or not the farmers themselves actually perceived any difference between the two extension approaches. It is possible that they have always used extension "lessons" as starting points for their own experimentation.

At most sites, there were multiple and contradictory claims in relation to the particular circumstances or persons associated with the introduction of widely recognized changes in farming practice. For example, at Kangare, the circumstances surrounding the introduction of napier grass as a cut-and-carry fodder crop are hotly disputed: Claims that it has been in the village for many years vie with claims that it was introduced more recently by the Livestock Department. Similarly, in Brong Ahafo, several individuals claim or were reported to have introduced key tomato varieties and other important changes in production practices.

Some of these apparently contradictory claims may be explained by the fact that whereas certain crops have been known for many years as small-scale or garden crops, they have only recently been transformed into commercial crops. This shift from a small-scale or garden crop to a commercial crop seems to be equivalent to the introduction of a whole new crop: Their respective histories are not readily linked in popular discourse.

At the same time, the contradictory claims may simply reflect the reality of "multiple sources" of innovation. In the case of stall feeding of dairy cattle in Kangare, for example, there have been many organizations involved, each distinguishable at some level, yet all promoting essentially the same set of technologies. Nevertheless, disputes concerning the origin of important innovations are certainly not unknown in the domains of formal science and industry, nor are instances of true simultaneous discovery. The study of these disputed claims among farmers might be a useful way of identifying particularly important innovations or events.

Over all sites, 83 percent of the examples of experimentation were classified as confirming or rediscovering something that was already known in the area, 7 percent were classified as resulting or potentially resulting in major modifications, and 10 percent as resulting in novel techniques, understanding, or models. Perhaps the most striking difference among the sites is that 31 percent of the examples from Kangare were classified as resulting or potentially resulting in something novel, while at the other sites, this outcome was associated with less that 13 percent of the examples. Kangare is also the site where the highest proportion of experiments had their origins in the farmers' own ideas.

The fact that over three-quarters of all examples of farmers' experiments were classified as resulting in the rediscovery of something that is commonly known is telling. The interviews point to the existence, in any particular place, of a more or less expansive repertoire of options for any specific operation within the crop production cycle. An individual may not be aware of those options that he or she does not normally use. But as circumstances change, as individuals move from mechanized to manual land preparation, for example, they use experiments to shift their farming practice within the bounds of the repertoire. The following examples from Zimbabwe illustrate this point:

> Normally she just covers sorghum seeds after broadcasting, but she has noticed that when the sun gets on the young plants they wilt. This year, rather than broadcast her sorghum, she wants to plant it on ridges. She has seen others do it, but she has never done it herself. She asked a few men about the practice and they said that the ridges were good because they increased the moisture in the soil. (z29)
>
> Normally she plants sorghum by walking behind the plow and dropping seed in the furrow. She has just planted a mixture of millet and sorghum by broadcasting it onto unplowed land and covering the seed with a hoe. This is the first time she has ever done this. This year "is a lesson," and she will see what happens. She says that almost all the women plant mixtures of sorghum and millet like this, but she did not talk to anyone before trying it. (z10)

The fact that many farmers' experiments seem to test practices and alternatives that are already generally known does not detract from the value of local experimentation. Rather, experimentation is an essential mechanism through which individuals overcome their own partial knowledge and limited experience. There is no alternative way for farmers to both access and integrate this information into their existing farming practice.

FARMERS' EXPERIMENTS IN EAST ANGLIA

In the United Kingdom, our examples of farmers' experiments came from the Ideas Competition database and from individual interviews. In relation specifically to machines and farm mechanization, the 334 entries in the Ideas Competition since 1948 provide a view of the topics around which farmers' experimentation and invention have clustered. Overall, the most common areas include entries relating to tractors, transport, and material handling (17 percent), land preparation (16 percent), spraying (11 percent), and livestock housing and handling (10 percent) (Figure 6.3). Together, these four topics account for 54 percent of contest entries since 1948. The fourteen entries for 1993 and 1994 included a modification to a moldboard plow, a set of soil packer rolls, a front-mounted cultivator, a furrow comb, a straw compactor, a sprayer, a press stabilizer, a wheel right, a muck barrow, a log splitter, a pig transporter, a calf transporter, and a fan for piglets.

Farmers in East Anglia also reported experiments relating to a broad range of other topics, from systems of pig breeding and tests of pig rations to herbicide application dates and rates. Many examples relate to the application (i.e., timing and rate) of agrochemicals including herbicides and fungicides (Table 6.3).

These examples paint a picture similar to the one that emerges from the Africa data. They are focused around a relatively limited number of topics, and farmers appear to be concerned primarily with small changes to a proven formula, using experiments to explore change at the margins of established practice. The many examples of experiments concerned with the rate and timing of herbicide and pesticide application can be seen in this light. The fact that entries in the competition are concentrated around a relatively limited number of very basic farm operations and problems is another indication of this process of experimentation at the margin of standard practice. Clearly, one major objective of farmers' experimentation relating to machinery is increasing the efficiency of machine use and operator time, as indicated by the relatively large number of contest entries that attempt to combine two or more operations such as tillage and planting or fertilizer application and planting.

This is not to say that farmers do not actively respond to new constraints and opportunities through experimentation. Indeed, from an analysis of the pattern of entries in the Ideas Competition, Knight (1995)

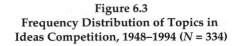

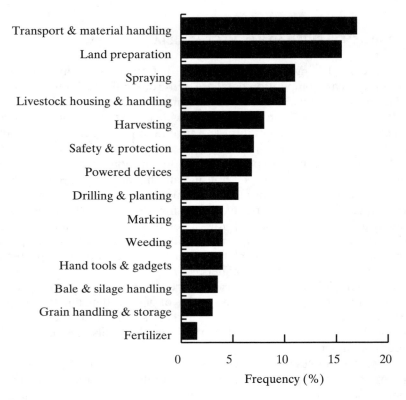

Figure 6.3
Frequency Distribution of Topics in
Ideas Competition, 1948–1994 (N = 334)

describes "waves" of interest focused on particular topics. The timing of these waves appears to be related to wider trends in the farming industry. For example, he cites the case of entries concerned with spraying equipment: There were no entries related to spraying prior to 1964, but during the 1970s and early 1980s, spray-related entries accounted for a significant proportion of all entries (Figure 6.4). This was also the period when the use of sprays, particularly herbicides and fungicides for cereal production, increased dramatically in East Anglia. Since the mid-1980s, however, contest entries focused on spray equipment have been much less common. Nevertheless, the central role of agrochemicals in East Anglian farming systems means that concerns with timing and rate of pesticide application, and the possibility of gains in efficiency and effectiveness with new tank mixes, continue to motivate much of farmers' experimentation (Lyon 1994).

Table 6.3.   Examples of Farmers' Experiments from East Anglia

Reference	Example
Lyon (1994:13)	"I test chemicals by cutting off half the (sprayer) boom for, say three years within a field. … Like this year I used a straw shortener. I wanted to know if it worked or if it was just the weather. … I noticed a 5–6 inch difference."
Rijal, Fitzgibbon, and Smith (1994:5)	Last season the farmer, apparently on a whim, used a corn hoe on a small section of his young sugar beet crop. Immediately afterward the crop looked decimated, but after a few days it responded very favourably. He says he will continue this method as standard practice, replacing the use of a beet hoe.
Lyon (1994:14)	"I had this problem with the company not giving me the mixture (of chemicals) early enough and they don't like to give it to you in the previous year because the mixture separates. … Anyway, I bought lots in the previous year and was vigilant with the mixing. … The company representative wouldn't believe it until I showed him. … I set up this trial last year with my method, the recommendation and no treatment. … I found a two to three hundred weight difference between no treatment and the recommendation and then a ten hundred weight difference between the recommendation and mine."

In terms of motivation, the examples from the United Kingdom illustrate both proactive and reactive experimentation. Although we have little specific data available, it seems likely that most machinery innovations that were eventually entered in the Ideas Competition would have had to have been planned to some degree. Similarly, the oft-reported example of leaving a small part of a field unsprayed is portrayed by farmers as a conscious decision that reflects both forethought and a certain degree of repetitive behavior. On the other hand, very clear examples of reactive experimentation were found. One farmer, for example, reported that, because the delivery of potato seed was late, there was no time for the usual practice of "chitting" (i.e., presprouting) the tubers before planting. The tubers were therefore planted directly as there was no other option. As the season progressed and the potato crop was obviously doing well, the farmer took increasing interest in what had become a "reactive" experiment.

Side-by-side comparisons also appear to be a common element of farmers' experiments in East Anglia, where, for example, farmers reported creating trials by turning off the herbicide spray boom in a small

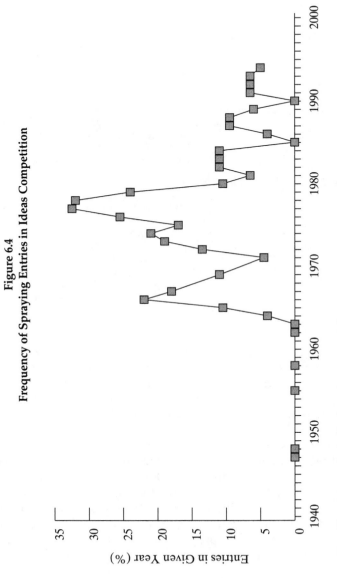

Figure 6.4
Frequency of Spraying Entries in Ideas Competition

Note: Figures were obtained using a three-year moving average.

part of the field. Another farmer reported that he regularly compares different pig feeds by varying the feed given to groups of pigs in adjacent pens (Rijal, Fitzgibbon, and Smith 1994; Lyon 1994).

In terms of the outputs of farmers' experimentation, it is interesting to note that Knight (1995:26) estimated that over the forty-six-year history of the Ideas Competition in Norfolk, approximately 75 percent of the entries could be characterized as "adaptations to existing equipment" as opposed to "inventions." This sense that many entries are, in effect, minor modifications to existing equipment is also reflected in comments by the contest judges. Originality is one of the criteria they use in awarding prizes, and in 1993, for example, their written comments about many entries stressed the lack of originality. However, in a situation where successful innovation has both financial and social implications, Knight (ibid.:37) noted disputed claims by farmers, on the one hand, and machinery manufacturers, on the other, as to the origin of certain ideas. Questions of origin and ownership can take on considerable importance as there are some examples of farmers' ideas that have been successfully commercialized, either by the farmers themselves or by machinery companies acting with or without the farmer's permission (Carr 1996). In this respect, it is interesting to note the contradictory attitudes of two machinery manufacturers toward the farmers' entries in the Ideas Competition: While the first states that "if it wasn't for farmers' ideas we'd all still be riding horses wouldn't we," the second suggests that "it's just a bit of fun isn't it, they all want to come up with the winning idea so they can be presented to the Duke" (Knight 1995:32–33).

SUMMARY

The examples from both Africa and the United Kingdom give us a particular, temporally determined view of what we assume to be a larger, continuous, and repetitive search for small gains through marginal changes in basic agronomic practices. The significance of the field data lies not in the particular distributions of topics associated with the examples of experimentation, but in the fact that the vast majority of examples revolve around a relatively limited array of standard and very practical agronomic concerns. Seen in this way, at least in terms of their subject focus, our examples of farmers' experiments have much in common with the experiments that form the backbone of formal applied and adaptive agronomic research. In common with farmers' experiments, much applied and adaptive research simply seeks incremental improvements through experiments to test minor modifications to an established combination of agronomic practices. For example, much formal agronomic research continues to be concerned with testing fertilizer and pesticide rates and times of application, varieties, inter- and intrarow spacings within the context of a given production system. Neither farmers nor applied agricultural researchers are concerned principally, or, we would argue, even to any significant degree, with either testing or contributing to theory, or determining the specific ecological, physiological, or biochemical mechanisms governing the outcomes of their experiments. Rather, they are both interested essentially in improving specific combinations of existing techniques and in expanding the catalog of available techniques and recommendations. To accomplish this, both farmers and agricultural researchers focus on the same, rather limited number of leverage points that are themselves determined by the very nature of crop production.

Our data indicate that the vast majority of farmers' experiments are proactive and that approximately half share key elements with standard agronomic trials: the use of side-by-side comparison and the inclusion of some kind of control treatment.

However, before continuing, we must ask whether these results concerning the motivation and methods associated with the examples of farmers' experimentation do not simply confirm the suggestion by Van der Ploeg (1990), Richards (1987), and others of increasing hegemony of the formal research systems over "indigenous" research approaches. This line of argument certainly does offer one explanation for the observed similarities between farmers' and agronomists' experiments. Yet it is a very real challenge to demonstrate that those indigenous research approaches and traditions would, in either form or interpretation, deal with, for example, a handful of "new" seed in a significantly different way. What would an indigenous, "culturally pure" experiment or test look like? In considering this question, Amanor (1993a:41) invokes the image of a "holistic approach which is concerned with synergism, systems dynamics, and the interrelationship of the energy cycles within the farm environment, rather than the isolation of single characteristics such as yield," but offers no concrete examples of such a process of indigenous experimentation. Although there is certainly abundant evidence that farmers are interested in crop characteristics other than yield, to our knowledge there is none to support the notion that they have developed conceptual or analytical frameworks and tools with which to grapple with concerns such as holism, synergy, energy flows, or system dynamics. Indeed, we have ourselves previously argued that farmers' analyses of new technologies are more encompassing and synthetic than can normally be handled in standard designs for agronomic trials (Sumberg and Okali 1988). Nevertheless, it is one thing to suggest that farmers may have multiple criteria for evaluation and a complex system by which these are weighted and synthesized, and quite another to speak of an ability to evaluate concepts such as holism and synergy. It is telling that the operationalization of these same concepts, to which we might well add "sustainability," continues to baffle those formal agricultural researchers who have themselves tried since the 1970s to move beyond simply narrowing the so-called "yield gap."

We understand that the passing of indigenous research approaches and traditions, if indeed that is what has happened, might be seen as an impoverishment of local culture and knowledge systems. However, it is critical to acknowledge that the methods and approaches that are so much in evidence in the examples of farmers' experimentation from Africa are probably of much greater value to contemporary farmers who seek to survive and move ahead in the face of changing personal, social, and economic circumstances. In situations where some inputs are purchased and crops are marketed, the choice between, for example, planting variety "A" or variety "B," or applying two bags of fertilizer per hectare or three, may well determine the difference between success and failure. Our data indicate that many farmers take a pragmatic and straightforward approach in their search for answers to such questions: They plant side-by-side plots over several seasons and observe. It is our view that the suggestion of declining local research traditions due to hegemony of formal science and the assumption of associated negative impacts on farmers have no basis in fact.

Fifteen percent of all examples of farmers' experiments from Africa involve the testing of a new crop, which implies at least the *potential* for fairly radical change in the farming systems if the experiments are successful. Apart from these, there are only very few examples of experiments that appear to be more radical, involving more complex changes to existing production systems or marketing arrangements. The bias toward relatively simple agronomic experiments may be a reflection of our research methods, but at the same time, it must be acknowledged that more complex experiments will pose greater difficulties in terms of management and interpretation, and they may well imply greater financial risk. Their relative scarcity among the examples should therefore not be altogether surprising.

However, it is critical to acknowledge as well that the image of farmers' experimentation that emerges from the data from both Africa and the United Kingdom is very much one of reinvention, fine-tuning, and tinkering. By and large, farmers seem to be using the experiments to make an existing production system or practice marginally better. While they appear to go about this task in a planned and somewhat systematic manner, relatively few farmers regularly use experiments to develop or explore novel ideas or techniques. The experiments they do and the manner in which they are done illustrates the importance to individual farmers of customizing or fine-tuning techniques and production systems to suit their individual circumstances.

7

Farmers, Experimentation, and Information

As highlighted through the review in Chapter 3, the literature on farmers' experiments provides no clear or consistent view of the type of person most likely to engage in experimentation. On the one hand, experimentation is portrayed as a natural, widespread, and essential element of farming; on the other, the use of terms such as "research-minded farmer" and "experimenting farmer" gives the impression that interest in, or skill at, experimentation is differentially distributed within the population. If the general proposition that all farmers must do experiments in order to survive is accepted, then there would appear to be little obvious value in categories such as "research-minded farmer" and "experimenting farmer." Or is the suggestion that, although all farmers experiment, a smaller group of research-minded farmers experiment more, or in a different way? We indicated in Chapter 3 that some of these contradictions can be explained by the lack of empirical data. In this chapter, we use our field research data to explore social, economic, and site-specific factors associated with farmers' experimentation. We also explore the links between farmers' experiments and local information networks.

SOCIOECONOMIC CHARACTERISTICS AND EXPERIMENTATION

Over all sites in Africa, 55 percent of individual interviews yielded at least one example of experimentation (Table 7.1). Nevertheless, the variation among sites was great, with as much as 83 percent of interviews in Brong Ahafo and as little as 27 percent in Eastern Region resulting in examples of experimentation. In general, respondents were more likely to report an experiment if they were either male, younger, more educated, had a resident spouse or were never married, considered farming their primary occupation, did not receive remittances, had been employed outside the

111

Table 7.1.   Distribution of Examples of Farmers' Experimentation

| Site | Interviews Conducted | | Examples Cataloged | | % of Interviews Yielding at Least 1 Example |
	No.	%	No.	%	
Kangare	14	7	16	10	71
Brong Ahafo	53	28	69	45	83
Eastern Region	73	39	27	17	27
Chivi	49	26	43	28	61
Total	189	100	155	100	55

area at some point, and had worked with the extension service or a project (Table 7.2). As indicated in Chapter 4, among the individuals we interviewed, these socioeconomic characteristics were not distributed independently: Male respondents who experimented generally had more education, were more likely to be married and have a resident spouse, to have worked outside the area, and to have worked with an extension or a development project.

Within the individual sites, the picture varies somewhat. In Chivi, those reporting experiments differed from the general profile given above only in that they were more likely to be older and to have less formal education compared to those not reporting an experiment; while in Eastern Region, neither farming as the primary occupation, nor nonreceipt of remittances, nor working outside the area was associated with the reporting of experiments. In Kangare, receiving remittances and not having worked with an extension organization had no effect on the reporting of experiments; while in Brong Ahafo, the reporting of experimentation was not associated with any of these particular socioeconomic variables.

Should we expect individuals with different socioeconomic profiles to do different kinds of experiments or experiments focused on different topics? Although the literature gives us little guidance in this matter, it certainly might be reasonable to expect, for example, that strong gender segregated roles within agriculture would orient men and women toward different topics for experimentation, or that differences in education and wealth might be associated with more planned or controlled experiments. We will first address these questions using the data combined over all sites, and then through an analysis of the disaggregated data from Chivi and Brong Ahafo.

Table 7.2.   Experimentation and Socioeconomic Variables, All Sites

Characteristic	No Example Reported	Example Reported	Number
Gender			
Male	36	64	106
Female	57	43	83
Age			
15–35	37	62	60
36–55	51	49	81
56+	46	54	48
Education (yrs.)			
0–4	58	42	97
5–8	41	59	56
8+	13	87	31
Marital status			
Never married	36	64	11
Divorced/widowed	51	49	45
Married	43	57	126
Resident spouse			
Yes	36	64	110
No	70	30	20
Primary occupation			
Farming	43	57	168
Other	62	38	21
Remittances received			
Yes	58	42	62
No	40	60	115
Worked outside area			
Yes	37	63	95
No	54	45	85
Worked with extension			
Yes	27	73	37
No	49	51	146

Using the 155 examples reported from 104 individual farmers from all sites, all the socioeconomic variables under consideration appear to be associated with differences in reported experiments. As indicated in Chapter 4, one must bear in mind that there are strong interrelationships among these socioeconomic variables: Compared to women, men are likely to be slightly older, have more education, are more likely to be married, and to have worked outside the area. The characteristics of the reported experiments that appear to be associated with differences in socioeconomic variables are motivation, method, source, outcome, and topic (Tables 7.3, 7.4, and 7.5).

Table 7.3.   Experimentation and Gender (% within gender)

Characteristic	Male	Female	Overall
Motivation			
Proactive	89	77	85
Reactive	11	23	15
$n =$	105	44	149
Method			
Without control	61	61	61
With control	39	39	39
$n =$	84	26	120
Source			
Trying what has been			
observed or suggested	48	59	51
Trying something that has			
been actively promoted	15	16	15
Own idea	38	25	34
$n =$	88	32	120
Outcome			
Something novel	13	2	10
Rediscovery; minor			
modification	79	96	84
Major modification	8	2	7
$n =$	109	45	154

Overall, experiments reported by men are somewhat more likely to be proactive than those reported by women, as are those reported by married individuals compared to widows and divorcees, by individuals whose primary occupation is farming compared to those who have different primary occupations, and by individuals who have worked outside the area compared to those who have not. In relation to the method of experimentation, examples reported by individuals who receive remittances are more likely to be without an obvious control plot compared with those reported by individuals who do not receive remittances.

In general, compared to those of older farmers, the experiments reported by younger farmers are more likely to be based on copying and less likely to involve trying something that is being actively promoted, while examples reported by those who have worked for an extension organization are less likely to be based on copying and more likely to reflect their own ideas. Experiments reported by men are more likely to result in outcomes that are novel and that represent major modifications, while those reported by women are more likely to result in minor modifications to existing practices. By the same token, compared to

Table 7.4.   Experimentation and Age (% within age ranges)

Characteristic	Age			Overall
	15–35	36–55	56+	
Motivation				
Proactive	82	93	78	85
Reactive	18	7	22	15
n =	55	57	37	149
Method				
Without control	56	57	74	61
With control	44	54	26	40
n =	43	46	31	120
Source				
Trying what has been observed or suggested	60	51	39	51
Trying something that has been actively promoted	5	23	16	15
Own idea	36	26	45	34
n =	42	47	31	120
Outcome				
Something novel	5	7	22	10
Rediscovery; minor modification	91	82	76	84
Major modification	4	12	3	7
n =	57	60	37	154

those of younger farmers, examples given by older farmers are more likely to result in something novel, as are examples reported by individuals who have worked for an extension organization and by those who have worked outside the area. Increasing levels of formal education are associated with reporting a lower frequency of experiments focused on new crop varieties, land preparation, and seeding method, and a higher frequency of experiments focused on fertilizer and soil fertility.

Moving now to the Chivi site in Zimbabwe, examples reported by men compared to those reported by women are more likely to be proactive, less likely to involve trying something that is being actively promoted, and less likely to result in only minor modifications to existing practices. Along these same lines, compared to those of younger farmers, examples by older farmers are more likely to result in a novel technique, while those by widows and divorcees are more likely to involve copying or trying something that is being promoted. Individuals who have

Table 7.5. **Experimentation and Education (% within education divisions)**

Characteristic	Years of Education			Overall
	0–4	5–8	8+	
Motivation				
Proactive	86	80	90	85
Reactive	14	21	10	15
n =	51	44	51	149
Method				
Without control	63	66	58	61
With control	37	34	43	39
n =	41	35	40	120
Source				
Trying what has been observed or suggested	45	48	57	50
Trying something that has been actively promoted	21	13	11	15
Own idea	33	39	32	34
n =	42	31	44	120
Outcome				
Something novel	9	11	10	10
Rediscovery; minor modification	87	80	82	83
Major modification	4	9	8	7
n =	53	46	51	150

worked outside the area are more likely to use a control plot, while those who have worked for an extension organization are more likely to do proactive experiments and to try their own ideas.

In Brong Ahafo, examples of experimentation reported by younger people are more likely to be proactive than those by older farmers, as are those by individuals who have worked outside the region. Examples from individuals who have worked for an extension organization are more likely to be without an obvious control plot and are more likely to be focused on new crops.

Overall, few clear trends emerge from these analyses. Perhaps the most that can be said is that there is some indication that men (associated with higher education and having worked for an extension organization) are more likely to report examples of experimentation than are women, and these examples are more likely to be proactive and to not involve copying or trying an idea that has been promoted. However, these associations are weak at best; perhaps more significant, then, is that examples of experimentation are reported by farmers across the socioeconomic spectrum and that no particular type of experiment appears to be exclusively associated with a single category of farmers. Neither gender, nor age, nor education appears to exert an overriding influence on the character of the examples.

At this point, it is interesting to speculate on the significance of gender roles in agriculture and the position of women vis-à-vis farmers' experimentation. Beginning with the work of Boserup (1970), there is a long line of studies that both portrays clear gender roles in African agriculture and locates rural women firmly in the traditional sector (e.g., Bryceson 1995a). The argument is that in many situations men have moved forward into commercial farming with the use of modern technology, such as new varieties, fertilizers, herbicides, and mechanization, while women have not (Stamp 1989). Accounts that deviate from this general scenario are few and far between. Bowen (1993) provides one of the few when he compares irrigated commercial gardens in Swaziland, where chemicals and modern varieties are used, with homestead lands, where maize is apparently grown as it has been for years. What is interesting is that both men and women control irrigated gardens, but only women work on the homestead lands. In this case, therefore, it appears that the use of innovations, including irrigation, new vegetable varieties, and chemical inputs, is not linked with gender. Idowu and Guyer (1991), working among the Yoruba of Nigeria, also report examples of women who have invested in commercial farming in response to an accessible and growing market.

Certainly, men's and women's knowledge varies, reflecting varied "interests," differential access to markets and resources, and socially defined gender roles. These differences in men's and women's knowledge and experience can have important implications for agricultural research and extension, as illustrated, for example, by Sperling, Loevinsohn, and Ntabomvura (1993), who used female "bean experts" in a formal bean breeding program. On the other hand, Coughenour and Nazhat (1985) compared male and female farmers in terms of their knowledge and use of certain innovations, their sources of information, and their speed of adoption, and they concluded that the main difference between male farmers and female farmers (who are also in a decisionmaking position with respect to farming) is that the females have fewer resources. Otherwise, women mention the same number of innovations as do men and adopt, even if somewhat more slowly, the same innovations. This is all very logical says Goldman (1991), who, among others, notes that subsistence production is not inevitably tied to traditional technology and that market production does not necessarily call for modern technology to the exclusion of traditional practice (although this is, in fact, implicit in much of the literature).

Based on reviews of women's access to extension services in different countries (Saito and Spurling 1992) and studies of women's roles and decisionmaking in agriculture (e.g., Spring 1988), women are now viewed by many rural service institutions as being underserviced and having special needs. Whether women have less access to local, informal sources of information relating to agriculture is not as widely discussed. Women obviously have access to some information, and it has been argued by McCorkle, Brandsletter, and McClure (1988) that, relative to the activities that are important to them, they get the information they need. So, for example, information about food processing moves freely among women in Niger. The fact that women spend much of their time with other women rather than men suggests that they might have limited access to information that was passed to men in the first instance, but women in Niger indicated to McCorkle's group (ibid.) that they got most of their production information from their spouses or other male relatives. Regarding innovations, women were unable or unwilling to identify innovations in agricultural production that they had developed or adopted, but they did report innovations relating to cooking stoves. McCorkle, Brandsletter, and McClure attribute this difference to gender-based role differentiation.

Thus, if, for example, gender roles are clearly defined and strictly respected, one might well expect there to be some discernible impact on experimentation. Roles defined by task or crop should, at the very least, affect the topic of the experiments being carried out by men and women. At the same time, a gender division of labor that results in women often working on fields and crops managed and controlled by their husbands might logically be expected to have an impact on their propensity to experiment and on the characteristics of their experiments.

Our data do not appear to support these expectations. Part of the explanation for this may be that there has been a progressive breaking down of strict gender roles within agriculture (Bryceson 1995b) or, alternatively, that these roles were never as set as has been portrayed in much of the literature (Stone, Stone, and Netting 1995). It is obvious, for example, that gender roles within agriculture must be affected in those areas with relatively high levels of either short- or longer-term male out-migration. The diversified, flexible, and responsive livelihood strategies that characterize so much of rural Africa are just not compatible with a strict gender division of labor. As to the effect on experimentation of an individual's relationship to a field or crop—for example, as owner or laborer—we are not able to comment directly.

Our research provides some tentative indications that members of more vulnerable groups, such as widows and the elderly, may be less able to successfully complete experiments that they initiate because of their limited degree of control over essential resources. A female household head in Chivi, for example, reported trying an application of manure on maize at the beginning of one year but not being able to repeat the trial the following season because she had no access to draft animals then; consequently she learned nothing of value (z36). In effect, experiments by vulnerable people are apt to be "overtaken by events." Thus, for these individuals, hazard is not only a factor promoting experimentation, but perhaps more important, it is one that actually limits it.

In this context, it is interesting to return to the contradictory claims highlighted in Chapter 3 concerning the relationship between poverty and farmers' experimentation. On the one hand, it is suggested that poverty forces people to experiment to ensure their survival; on the other, it is posited that poverty severely inhibits experimental or innovative behavior. At the root of this apparent contradiction is the distinction between a need or interest in experimentation and the ability to carry through an experiment. It is obvious that experiments of certain types and scales will require levels of resources that are out of the reach of some people. Nevertheless, from the nature and scale of the majority of our examples, we conclude that it is not necessarily the quantity of resources or the associated risk that limits experimentation by members of marginal groups; rather, it is a lack of control over those resources that allows experiments to be "overtaken by events."

INNOVATORS AND RESEARCH-MINDED FARMERS

One question that arises at this point is whether those persons who report examples of experimentation are essentially the same persons described by Rogers (1960) and by others within the diffusion of innovations tradition as "innovators." In terms of the process of technological change, innovators have been defined as the first 2.5 percent of the population to adopt an innovation. These farmers are characterized as having "higher education, larger farms, higher incomes, higher social status, and wider travel than the average farmer" (ibid.:409). They are also "cosmopolites rather than localites," young, "research-minded," and often belong to formal organizations outside the community.

Given that in three of our four sites, more than 60 percent of the individuals we interviewed reported at least one example of an experiment, it seems unlikely that they are the same as the innovators defined by Rogers. It must also be remembered that those studying the diffusion of agricultural innovations most often focused on fairly major changes in farming practice as indicated by the implementation of a decision to "adopt" a particular innovation. At this level, the changes being studied are usually well defined, more often than not actively promoted by an extension service, and probably place significant new demands on resources. At least initially, the diffusion-of-innovations school ignored the smaller-scale testing that was eventually recognized and legitimized as "reinvention." We now understand that this testing—in our case, in the form of farmers' experiments—is only a relatively small part of the process leading to a decision of whether or not to adopt. Thus, if they result in any change at all, farmers' experiments or reinvention can also yield less dramatic change, in the sense of adaptation rather than adoption. And there are clear implications of a focus on full-scale adoption as opposed to small-scale "trying," in terms of, for example, our understanding of characteristics of "innovators" or "farmers who experiment" and the importance to the farmer of resource availability and financial and social risk.

If those reporting examples of experimentation are not Rogers's "innovators," can we now say anything about the notion that there is a subgroup of research-minded farmers who, through their involvement with experimentation, are somehow different from the population of farmers as a whole? Are there some individuals who do more experiments, or who do experiments that are consistently different in terms of topic, method, and so forth?

During the period of our field research, we certainly met a small number of individuals for whom experimentation appeared to play a somewhat more significant part in their farming activities. One individual in Brong Ahafo, for example, showed us what he described as "an experiment farm" (g222). At the time of the interview, he had experiments in progress on planting density and the use of liquid fertilizer when transplanting tomato seedlings. He specifically invited his friends to the farm when the liquid fertilizer was being applied, and he pointed out during the interview that although some people do experiments for selfish reasons, he tries to share the experience and the results with others. He cautioned us against judging the results of an experiment too quickly and moving too quickly to cover the whole farm with an innovation, for "if good can come [from an experiment], bad can come too." He is thirty-nine years old, has nine years of formal education, controls a large area of farmland, and has an uncle that works for the Ghana Cocoa Board. He portrays himself as a dedicated, full-time farmer who has never done other work: "Farming is my background, my backbone."

At some of the other sites, a limited number of similar individuals were identified, and we have certainly met others over the years. However, in the light of the fact that over half of the interviews resulted in evidence of experimentation, we are very hesitant to conclude that these few individuals represent an important subpopulation of farmers who are more engaged in experimentation. They may well be more articulate, more organized, or more public-spirited, but we have no evidence that they are fundamentally different in their behavior or their understanding of their own experimental activities. As will become evident in later sections, this conclusion is supported by the observation that at none of the sites were respondents willing or able to identify other individuals who were generally recognized for trying new things or experimenting.

INFLUENCE OF THE SITE

In addition to socioeconomic and personal characteristics relating to individual farmers, we identified a number of site-related factors that appear to have some effect on farmers' experimentation (Table 7.6). These factors are discussed in detail below, but before addressing them, it is important to note that they are clearly not independent and that the interactions among them are likely to be complex. It is also likely that at any given site, a number of these factors will be at work; thus the precise contribution or effect of any individual factor cannot be isolated or estimated. Nevertheless, at least in a preliminary way, the role and importance of these factors can be illustrated with reference to the research sites.

Perception of a Problem

Many of the examples of experimentation indicate that they are proactive, planned activities. Farmers' experiments seem to cluster around a limited number of central themes where there is a shared perception of a problem or problems that can possibly be addressed, at least in part, through individual management decisions. Thus, the difficulties associated with recurrent drought in Chivi have done much to focus the minds of farmers there on ways of increasing soil-water availability and alternative methods of land preparation following the loss of draft animals. Similarly, a general perception of declining soil fertility in Brong Ahafo appears to be the stimulus behind many farmers' experiments in that region. At Kangare and Boma on the other hand, there appeared to be no clearly identified or widely acknowledged problems. Elsewhere, where problems are not readily amenable to individual intervention, such as the generally cited issue of low tomato prices in Suminakese, individual experimentation seems to be less widespread (although in Suminakese there was a group-based attempt to solve this particular problem).

Table 7.6.   Factors Associated with Farmers' Experimentation

Factor	Limited, Unfocused Experimentation $\longrightarrow$	Abundant, Focused Experimentation
Perception of a problem	weak	strong
Commitment to farming	low	high
Transparency of the landscape	forest	savanna
Crop propagation method	vegetative	seed
Life cycle of crop	perennial	annual
Socioeconomic position	vulnerable	not vulnerable
Promotional activities	absent	present
Commercialization	low	high

Commitment to Farming

Diversification is one of the central characteristics of rural livelihood strategies in Africa. In addition to a wide range of agriculture and natural resource–based activities, rural and urban wage employment make critical contributions to many rural livelihoods. One result of this diversification, and the associated temporal and spatial patterns of migration and residence, is that there are many people in rural Africa who are engaged in agriculture as a secondary or even tertiary activity. While at any given site there is often much variation among individuals in terms of relative emphasis on particular economic activities, because of geographic location, natural resource endowment, tradition, and so forth, sites take on certain general characteristics vis-à-vis the level or type of engagement in agriculture. For example, the Luo people of western Kenya have a long tradition of investment in education, and Luo men have for many years migrated in search of wage labor. As a result, rural villages are populated largely by older people, women, and children, for whom farming is either an activity of last resort, a secondary occupation, or essentially an unpleasant and unavoidable accompaniment of rural residence. The situation at the sites in Eastern Region is somewhat similar. In these circumstances, it is hardly surprising that farmers' experiments appear to be unfocused (Kangare) and taking place at a relatively low level (Eastern Region).

In contrast, at Chivi and Pamdu, a significant proportion of the resident rural population appears to rely more directly on farming to sustain its livelihood and has a commitment to farming as a "chosen profession." This was particularly striking in Brong Ahafo among the group of relatively highly educated young men who are aggressively and successfully exploiting the income-generating potential of vegetable production. A number of our respondents reported that they first got involved in growing tomatoes when they were forced to return to the village because they could not afford to purchase tools after completing their formal education and an apprenticeship in carpentry or plumbing. However, the significant returns that they now gain through vegetable production lets them, with quite apparent pride, describe themselves first and foremost as farmers. In these circumstances, where farming is the central element of people's livelihoods, experimentation is both common and focused. An obvious self-reliant attitude was also evident among informants in Brong Ahafo (e.g., "Nobody knows more than somebody, you trust yourself" [g235] and, "There is nothing that will come that I do not know, I can treat anything" [g259]). Their self-confidence seemed to be based on a recognition of their own ability to experiment and solve problems, in contrast to our respondents in Kangare, who assumed that all relevant information was already at hand.

Transparency of the Landscape

Farming can be a more or less public performance depending, in part, on population density of the area and the nature of the landscape within which it takes place. The importance of simple observation in relation to farmers' experiments is illustrated by the fact that, overall, the source of the idea for more that 30 percent of the examples was classified as "trying something that had been observed or suggested." More open, savanna landscapes, as in Brong Ahafo and Chivi, facilitate the movement of information through simple observation. For example, in Chivi, in order to arrive at the field of one of our interviewees (z201) where we planned to conduct our interview, we had to walk with him on a path that crossed a large, open area of intensive cultivation. Without any prompting, he gave us twenty minutes of running commentary about the fields and crops that we passed. Private farming or private experimentation would appear to be a near impossibility in this context.

On the other hand, the combination of remnant forest, cocoa plantations, and dense stands of cassava at some sites in Eastern Region, for example, makes it almost impossible to develop a sense of ongoing farming activities, or even to determine whether or not someone is in the immediate vicinity. It is clear, however, that within a particular site and landscape, there are farmed areas that are more and less transparent. At Pamdu in Brong Ahafo, for instance, tomatoes are produced during the dry season in two distinct situations. The valley of the Odum River to the west of the main road is open, almost completely planted with tomatoes, and the activities of many individual farmers are easily observed. In con-

trast, to the east of the road, tomatoes are produced in smaller, more isolated plots cleared from the dense riverine vegetation.

The interactive effect of these site-related factors on levels of farmers' experimentation is suggested by the case of tomato production in
Suminakese. Here, although tomato production is taking place in a generally open landscape, the fact that it is a long-established activity and that
prices are now considered to be extraordinarily low may account for the
observation that tomato production is not a focus for experimentation.

Crop Propagation Method

Vegetative propagation of crops eliminates the need to go through a cycle of sexual reproduction in order to plant a new field. While vegetative reproduction offers many direct advantages to farmers, the elimination of the sexual cycle means that the opportunity for the creation of
new combinations of genetic material is greatly reduced. Crops that are
vegetatively propagated, such as cassava and plantain in Eastern Region and sugar cane in Kangare, provide less opportunity for experimentation as new genetic variability is not produced and handled with
each crop cycle. This limits the opportunity for selection and testing on
the part of the farmer. The significance of this is clear when we remember that the testing of new varieties was the second most common topic
of experimentation over all the sites in Africa. It is also worth noting
that the testing and development of crop genetic material is probably
the most commonly cited activity in the broader literature on farmers'
experiments.

The lack of experimentation associated with vegetatively propagated crops was even apparent in vegetative crops locally defined as
commercial, such as onions produced from sets in Eastern Region.

Life Cycle of Crops

We have already suggested that, to a large degree, the very nature of
crop production circumscribes the domain of farmers' experimentation.
The cycle of land preparation, crop and variety selection, planting, weeding, and so on, provides the framework within which a limited number of
key decisions, around which experiments might be contemplated, are
made. Perennial or long-cycle crops, such as cocoa, plantain, and oil palm
in Eastern Region and sugarcane in Kangare, therefore offer relatively
fewer clear points for experimentation than do annual crops, as decisions relating to variety, spacing, land preparation, and other considerations are made far less frequently. At the same time, a particular aspect
of the performance of a perennial crop, as observed at any given point in

its cycle, may reflect, to a greater or lesser extent, the accumulated effects of decisions made and conditions experienced several years earlier. Experiments with perennial and long-cycle crops thus pose additional challenges in terms of the interpretation of cause and effect, challenges faced by farmers and agricultural researchers alike. Nevertheless, a number of examples of farmers' experiments were reported relating to the nursery stage of tree crops, and a farmer in Kangare reported experimenting with a pest control method for sugarcane.

Promotional Activities

As indicated earlier, agricultural extension activities in Africa are commonly described as being somewhere between ineffective and seriously misguided. There is little question that many extension programs, both governmental and nongovernmental, have attempted to be highly directive and, in effect, to restrict rather than foster reinvention of the technologies being promoted. It is now widely recognized, however, that even without official encouragement, farmers deconstruct complex technical packages and adapt individual components or combinations of components as they see fit. We have previously explored the implications of this process of deconstruction, or the unpacking of a technical package, for formal research and extension (Okali, Sumberg, and Reddy 1994). Even if this process is not explicitly among its objectives, an extension presence, by providing new ideas and materials, may well stimulate farmers' experimentation rather than block it, as is implied in some discussions.

We should not be surprised when it appears that extension activities that promote the testing of new technologies, such as ITDG's extension of conservation tillage techniques in Chivi, have an impact on farmers' experimentation. Many farmers in Chivi referred directly to ITDG personnel or activities in describing the origin of their own experiments. The presence of ITDG is of interest not simply because it promotes a certain set of technologies, but perhaps more generally because it promotes individual and group experimentation as a legitimate element of the technology transfer process, regardless of the particular topic or technology. The Farmers Clubs in Chivi play a similar role, acting as formal channels for the transfer of information and ideas that apparently influence the propensity to experiment. However, it is important to note that the absence of formal promotion activities in other sites did not mean that farmers were not trying to resolve perceived problems at these sites. For example, the fact that there is little formal research or extension emphasis on vegetables in Brong Ahafo may have increased, or at least determined the emphasis of, the farmers' experiments there.

Level of Commercialization

The twin aspects of risk and competition that accompany increasing commercialization—aspects that are especially associated with high-value, perishable crops such as vegetables—appear to be important stimuli to experimentation. This was well illustrated in Brong Ahafo, where the particularly vibrant market for tomatoes has created an environment that appears to be conducive to very active experimentation. This site contrasts with Suminakese, where producers expressed to us dissatisfaction with the tomato market and did not demonstrate an active interest in trying new techniques or approaches. Indeed, it could be argued that increased commercialization not only stimulates experimentation, but also that increased commercialization is less likely without experimentation.

LINKS BETWEEN FARMERS' EXPERIMENTS
AND LOCAL INFORMATION NETWORKS

Our interest in local information networks relates to a desire to evaluate the notion that farmers' experiments have a potential development value that goes beyond the direct benefits to the individual experimenting farmers. This notion is an important element of the participatory research story line as outlined in Chapter 1 and the synergy hypothesis as introduced in Chapter 2.

To start, it is reasonable to assume that if the information gained through farmers' experiments has a wider value, it will tend to move through local information networks. There are certainly examples that demonstrate that technologies or practices with obvious substantial benefits diffuse almost automatically, although the rate and extent of diffusion will likely depend on issues such as the complexities of the innovation or the availability of required inputs. It also seems plausible to assume that if there are individuals who are particularly involved in experimentation, who experiment significantly more or more effectively than others, and might therefore be considered "research-minded farmers," then they will be generally known as such within their local communities.

In order to examine these propositions, farmers were asked to name people they knew who tried new things or who were considered locally as "experts." A second exercise entailed naming particularly well-informed individuals, or other sources of information, that the informants themselves use or would use if they needed help with a particular crop-related problem (i.e., "To whom or where do you go for advice if you see a new disease on your tomato plants?"). In fact, this approach is similar to that taken by diffusion researchers in their attempts to identify "innovators" and "opinion leaders" within agricultural communities. However, our interest was to determine whether farmers differentiate among themselves in terms of their propensity to experiment (just one specific component of the behavior of an "innovator" according to Rogers [1960]), whether certain individuals are generally recognized as experimenters, and whether individuals recognized as experimenters are also recognized as good sources of information or advice.

Before moving to a discussion of the results of these exercises, it is important to note that there were considerable differences among sites in the extent to which respondents were willing to give us the names of other persons when we posed questions about people who try new or different techniques or who are good sources of information. Specifically, in Chivi and Brong Ahafo, respondents were generally willing to engage in this type of discussion, but in Kangare and Eastern Region, they were considerably more hesitant. At these latter sites, common responses included: "How would I know what other people are doing?"; "I don't go around looking at other people's farms"; "No one here is doing anything different"; "Everybody knows how to grow it."

There are probably a number of factors that are associated with these differences in willingness to name names and discuss the innovative behavior of others. Certainly, the site-related factors discussed above, and especially the visibility of farms and plots, must come into play. But two other factors should be highlighted. The first relates to the nature of the communities within which our research was carried out. All of the sites were established communities where either most of the individuals interviewed or their kin had lived for many years. And although curiosity may be recognized as valuable by these communities, it is, at the same time, seen to be almost indistinguishable from gossip. At the Awisa site in Brong Ahafo, for example, visiting someone's farm without permission is regarded by some as highly suspicious and associated with bad luck and magic (i.e., juju). Gossip is seen as particularly suspicious and dangerous if strangers (i.e., anyone not born in the communities in which they are living) are involved, which was more likely to be the case in Kangare, Chivi, and Brong Ahafo than in Eastern Region.

Cohen (1993) provides another possible explanation in this context of the nature of communities for the apparent unwillingness to discuss innovative behavior. He suggests that although community members appreciate differences of opinion and knowledge in discussions among themselves, in discussions with outsiders, they might present their knowledge as monolithic and homogeneous to adhere to "the imperatives of egalitarianism and community solidarity." Particularly in both Kangare and Boma, we were told to "ask anyone and they will tell you; if someone refuses to tell you, come to me and I will tell you." We repeatedly heard phrases such as, "We are all doing the same thing here," "I do what my mother did, I learnt it from my mother," or "I don't do anything different, everyone here knows and does the same thing." Cohen suggests that such general statements do not compromise an individual's interests and that this "common knowledge" may simply be a "masquerade" or a "symbolic marker" that separates, in a conceptual sense, a community from the rest of the world. These are the normative concerns that all researchers must address in either short- or long-term research. Another possible example of community solidarity in the face of the outside world was the naming of former extension staff as key and knowledgeable individuals. Perhaps they were named simply because of an assumed rightness about their position, a rightness that outsiders such as ourselves would appreciate. An alternative explanation, of course, is that this kind of common knowledge is just as it appears: obvious, uncontested, and consisting of widely accepted facts, methods, and explanations.

The second factor that may be associated with the apparent hesitancy of our respondents to discuss the names of more innovative or well-informed farmers relates to the dominance of old, established crops and cropping practices in a number of the sites and a lack of a perceived need to change these. At the sites in Eastern Region, for example, farming is still dominated by low-input cocoa production along with cassava, plantain, and maize. Although government extension and research staff express concern over various problems, such as the disease black Sigatoka (*Mycosphaerella fijiensis*) in plantain, the farmers themselves seemed unconcerned. Similarly, at Kangare, maize is a staple crop of long standing, which continues to be produced successfully with few apparent problems. Even in Suminakese, the techniques of upland tomato production have not changed substantially over the years except for the introduction of agrochemicals. The one major problem area identified in this site, marketing of tomatoes, was being addressed by one man whose efforts were widely acknowledged.

When they were willing to name names, both men and women, in general, tended to name men: Few women were identified through these exercises at any of the sites. Certainly, as expected, relatives were widely reported in all sites as a source of both information and planting material. If pressed, those respondents who were initially hesitant to identify anyone might name a kinsman, especially if that kinsman was also present at the interview. And in some instances they actually named themselves! This naming of kinsmen might be related to an expectation that our work would lead to a development project or some other intervention, from which named individuals might then benefit. In both Kangare and Eastern Region, informants constantly queried our presence and purpose. Kangare has a history of many agricultural projects, and some of the more recent projects have included the free distribution of substantial material benefits, including dairy cows. In Eastern Region, particularly at the Boma site, farm visits are associated with government compensation schemes for diseased cocoa trees or for trees that have been destroyed by fire.

At these sites, and at the other sites to a more limited degree, we were repeatedly struck by the pervasiveness of this "project mentality." It presents agricultural researchers and development workers with a serious challenge, as the expectation of benefits colors all interaction with local people. This is particularly important if, as in the present case, the goal is to explore ideas as personal and ephemeral as farmers' experiments and local information networks. The project mentality should be recognized as a potential and increasingly important constraint to agricultural research and development programs that do not necessarily have the time or resources for long-term, in-depth studies. It is of course possible that long-term, in-depth studies involving no obvious benefit to participants may also be obsolete in this context. At the same time, there is little indication that the research and communication tools and techniques that are now commonly associated with rapid and participatory rural appraisal offer a solution to this particular conundrum.

Only in Brong Ahafo and Chivi were respondents generally willing to name other individuals. At the former site, there is a cohort of educated, commercially oriented young people who are actively engaged in vegetable production with the express purpose of generating cash income. While there is an almost palpable sense of competition within this cohort, the commonalties among the young producers and the fact that individual success depends to a significant degree on the village being known as a major and dependable center of production appear to support information exchange and communication. In Pamdu village, for example, one of the vegetable marketing groups held weekly meetings during which the techniques of tomato production were discussed. One informant (g261) indicated that when he first began to grow tomatoes, he attended these meetings but never officially became a member of the group. In Chivi, on the other hand, our research was actually carried out under the auspices of ITDG's ongoing development project, which is itself trying to promote local experimentation and information exchange. We suggest that the shared sense of challenge in the face of a series of poor harvests and a very uncertain beginning to the agricultural year, in combination with the cohesive role of the Farmers Clubs, helped to facilitate the naming exercises at these particular sites.

Perhaps the most significant finding of these naming exercises is that over all sites, few if any individuals were identified by our informants as being particularly engaged in experimentation (in the sense of trying new things). Thus, neither our own interviews nor the naming exercises identified individuals who reported themselves as, or were locally perceived to be, particularly innovative or experimental in the context of their farming activities. We do not, however, take these results to mean that there are no particularly innovative individuals (as opposed to individuals who do experiments) at the study sites. Rather, our conclusion is simply that the notion of an identifiable subgroup of research-minded farmers or farmer innovators, who distinguish themselves from the general farming population by their greater involvement in or skill at experimentation, is probably of little value.

Those individuals who were named by our respondents can be grouped into four categories (Table 7.7), although it goes without saying that the boundaries between these are not sharp, and some individuals fall into more than one category.

The first category is the "local subject matter specialist," whose special knowledge of, or interest in, an area was clearly identified with experience gained from practice rather than formal training. In addition, it appeared that these individuals successfully combine a variety of information sources to inform their own production operations. A typical individual identified in this category was a widower in Kangare (k17), who was widely identified as a "tomato expert." He described himself as "always picking up seed and trying different things," but also pointed out that he follows the advice of the ministry to rotate the tomato plot every two years. He also demonstrated his own "innovation," a small "cap" made from eucalyptus leaves and twigs to protect young tomato seedlings from the sun immediately after transplanting. The story of his success with tomatoes illustrates the importance of combining information from a number of sources: (1) experience—he knew how to handle small-seeded crops and how to make and manage a seedling nursery; (2) the formal system—after his first year planting tomatoes, he sought information from an extension agent about the pruning of suckers; and (3) his own experiments—he developed the shade cap to protect the seedlings after transplanting.

Although this farmer is considered a tomato expert, and other farmers in Kangare were aware of the shade cap he had developed, his cap was considered by some to be of limited value because of the time required to make it. We were informed that because the tomato expert has only a small plot, he can afford to make a shade cap for each plant. This example illustrates the very individual nature and situation-specific value of the outputs of at least some farmers' experiments. Thus it appears that the community recognized this individual's general understanding of tomato production, but saw little within his experimental

Table 7.7.   Categories of Named Individuals

Category	Characteristics
Local subject matter specialist	Usually male; recognized as having a particular interest in, and knowledge of, an area such as fruit production, herbal medicine, etc.; no formal training
Technical agent	Male; some formal technical training; was or is associated with extension service, research station, or development program; resident or retired in village; has links to technical services, uses recommended techniques, and has access to inputs; potential information gatekeeper
Successful farmer	Predominantly male; large-scale producer; likely to hold some official family, community, or church position(s); uses recommended techniques and has access to inputs; potential information gatekeeper
Prime mover	Male; likely to have traveled relatively widely; associated with a major change or introduction

activities of any particular or direct value. This point is further illustrated by the tomato expert's own observation that although he established a citrus orchard in order to test a wide range of species and varieties, the only reason other villagers come to visit is to eat the ripe fruit. They are not interested in information about his collection or in obtaining grafting material. A very similar observation was made by a local subject matter specialist in Eastern Region (g19). Other subject matter specialists were identified with interests in, for example, herbal medicines (Eastern Region), trees (Kangare), citrus fruit (Eastern Region), and tomato production (Brong Ahafo).

The second category is the "technical agent" or "professional agriculturist," who was closely associated in the naming exercises with the formal agricultural services. The prevalence of this type of person among those named may reflect how the informants assessed our own interest, but nevertheless, it was clear throughout the research that connections with formal agricultural research and extension programs were both widespread and valued. At every site, there was at least one individual who had worked within the extension service; and in Boma, up to one-third of all the individuals interviewed had some direct or indirect (e.g., through spouse) relationship with formal extension. In the case of the named individuals, the majority were locals who had worked as extension officers or research technicians with government or nongovernment agencies. Some were still employed by these organizations. It is interesting to note that in their own farming, these individuals appeared to be using the agronomic "packages" (line planting, improved variety, higher density, etc.) that have been so widely criticized in some parts of the literature, but that were seen by many respondents as being closely linked with the level of success enjoyed by these technical agents. It was not at all obvious from our interviews with them or other respondents that individuals named in this category were particularly concerned with experimenting with or modifying these technical recommendations.

Unlike the subject matter specialists, there is little doubt that individuals in the technical agent category are viewed as an important resource by other farmers in their local communities. In addition to their specific technical knowledge, they have an understanding of the structure and workings of the official bureaucracy and often retain some links with formal institutions that can be of significant value to their neighbors. In fact, we suggest that individuals in this category are widely distributed in rural areas in Africa and probably have some considerable influence on the processes of technical change within agriculture. While they have heretofore been largely neglected by formal agricultural research, extension, and development agencies, they may actually dwarf—in terms of number, quality of interaction with the farming population, and impact—the present staff of the formal agricultural extension systems.

The "successful farmer" category was the one on which most respondents initially focused. Indeed, a number of people indicated that it was really only this type of farmer—having more resources, farming at a larger scale, and selectively following extension recommendations—who was worth naming. Clearly, individuals named as subject matter specialists or technical agents may also be considered successful farmers within their communities. As noted above, many people in the study sites had formal training or links with extension, so those actually named as technical experts may also be recognized for their success in farming. It seems likely that successful farmers have benefited over the long term from closer relations with research and extension, and their recognized status thus reflects, at least in part, the success of these programs. Nevertheless, except for in Chivi and Brong Ahafo, the number of individuals named in this category was quite small.

The final category, the "prime mover," includes individuals who were associated with a major change in farming practice or a specific, discrete introduction of plant material or new technology. In a general sense, these individuals may be comparable to the "innovators" described by Rogers (1960). One man in Brong Ahafo, for example, was identified repeatedly for having introduced the Power variety of tomato, upon which the present intensive production system is said to rely. These individuals were named at all the sites, but in the case of Suminakese, a man who is attempting to orchestrate a fundamental reorganization of tomato production and marketing was the only individual named.

EXPERIMENTATION, INFORMATION TRANSFER, AND THE ELEMENTS OF SUCCESS

In Chivi and Brong Ahafo, because of the respondents' willingness to name names, the questions about different farmers and their roles in the local information network were investigated in more detail. We are certainly aware of Cohen's (1993) warning against taking the knowledge of local experts for granted and the fact that within communities there can be a number of versions of reality. However, in our work, we were not identifying experts in order to extract their version of reality: Our objective was simply to assess the possibility of identifying individuals and groups, and then to analyze the extent of agreement about these persons' expert status.

In stark contrast to respondents in Kangare and Eastern Region, people interviewed in Ward 21 in Chivi were generally very willing to cite and discuss the farming activities and farming styles of their neighbors. During the first few days of our fieldwork, through an iterative process of word identification and testing, two Shona words were identified that seemed to convey the idea of someone who is an innovator, dynamic, a go-getter: *mhare* and *changamuka*. We then used these words to ask: Who do you think are the most *mhare/changamuka* farmers in the local area? A total of fifty individuals (twenty-one men and twenty-nine women) responded to this question. Relatively few individuals were unable or unwilling to cite somebody. Of those who did not cite anybody, some indicated that nobody they knew deserved to be considered *mhara/changamuka,* as evidenced by the fact that everybody had received food aid following the recent poor harvest. Others indicated that they were simply not in a position to know enough about other people's farming activities to judge. In addition, questions were posed to bring forward information concerning the respondents' perceptions of the major differences between the named individuals and other farmers. Many of the individuals who were cited were also among those interviewed. Several others were subsequently interviewed simply because they were so frequently cited.

In total, thirty-five individuals were cited as being *mhare/changamuka.* Of these, the majority (65 percent) were cited only once or twice

(Figure 7.1), while four individuals accounted for 50 percent of all citations (Figure 7.2). All those cited by both men and women respondents were men: When this was called to the attention of two female respondents, they then named two women as *mhare/changamuka*. Unfortunately, we were unable to interview either of these women.

We were able to interview three of the four most commonly named men (z201, z235, and z55), however. Compared to other men in the sample, they were: slightly older; more likely to have had some past involvement

Figure 7.1
Frequency Distribution of Number of Times
Individuals Were Cited as *Mhare/Changamuka*

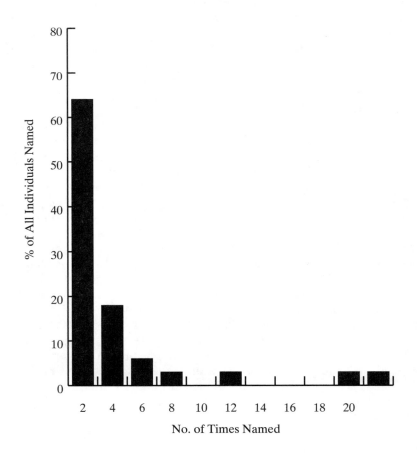

No. of Times Named

135

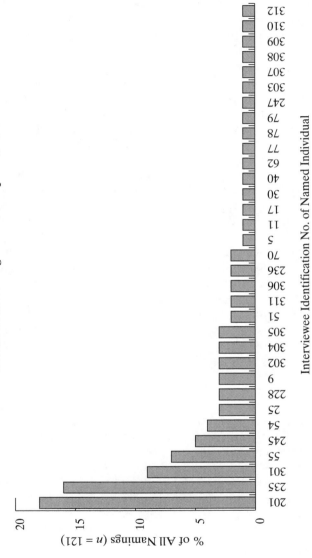

Figure 7.2
Frequency Distribution of Particular Individuals
Cited as *Mhare/Changamuka* (50 respondents)

(i.e., employment) with the agricultural extension service; less educated; slightly more likely to be a Farmers Club member; and much more likely to be a Farmers Club officer and to hold another position of status within the community. As shown in Table 7.8, three of the four most oft-named individuals played at least two other important public roles, and the most oft-cited individual played four of these public roles. This individual ($z201$) is commonly referred to by the nickname *"sadza"* (the staple maize meal) because "he always has enough to eat."

Only one person ($z55$) appeared to have achieved a level of general recognition for his agricultural activities independent of other major public roles (it is important to note, however, that he and his wife brew and sell sorghum beer on a significant scale at their home). There were no apparent differences between those most commonly cited as *mhare/changamuka* and other men in terms of previous employment outside the area or marital status.

On average, women named a smaller number of individuals as *mhare/changamuka* than did men (1.8 vs. 3.3). There is, in addition, some indication that women have a more narrow, or perhaps a more precisely defined, view of their innovative or dynamic neighbors. While men named twenty-five individuals as *mhare/changamuka*, only nineteen were cited by women. For women, the four most oft-named individuals accounted for 56 percent of citations (compared to 45 percent for men), while the top two accounted for 44 percent (compared to 26 percent for men).

The commonly identified indicators of the success of *mhare/changamuka* farmers were: (1) consistently having large grain harvests, as well as the ability to always, even in bad years, harvest something; (2) consistently having enough grain to eat, store, and sell; (3) consistently having crops that "look good"; and (4) being copied by others.

Table 7.8.   Roles Played by Oft-Named Individuals

	$z201$	$z235$	$z301$	$z55$
% of all citations	18	16	9	7
Public role:				
Kraal head	yes	yes	no	no
Preacher	yes	no	yes	no
Farmers Club chair	yes	no	yes	no
Master Farmer	yes	yes	yes	no

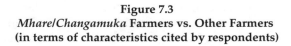

Figure 7.3
Mhare/Changamuka **Farmers vs. Other Farmers**
(in terms of characteristics cited by respondents)

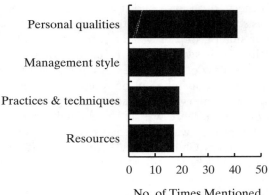

No. of Times Mentioned

When we asked our respondents about the difference between *mhare/changamuka* farmers and other farmers, their replies generally fell into four broad categories, with personal qualities such as "effort," "brains," and "attitude" being more commonly cited than either management style (planning, timeliness, constant work), or practices and techniques (early planting, deep plowing, correct spacing), or access to resources (implements, draft animals, inputs) (Figure 7.3). Overall, the responses pointed to clever, hardworking individuals with access to adequate implements and inputs. It is interesting to note that the idea of "innovativeness" and an individual's ability to adapt to situations (unless one assumes these traits to be inherent in the commonly cited words "brains" or "clever") were not closely associated with *mhare/changamuka* as they were cited only four times.

Thus, in Ward 21, when asked to name successful farmers, who are presumably key to local information networks, our respondents identified a wide range of individuals. However, the majority of citations clustered around a relatively restricted group of male, larger-scale farmers, who also play one or more other public roles. Most of those named repeatedly had received some Master Farmer training, and several had had some previous employment relationship with the agricultural extension service.

This points to the key role played by the Farmers Clubs and Master Farmers in information transfer and on people's perceptions of farming.

This role has been enhanced recently with the shift in the orientation of some Farmers Clubs from rather elitist groups restricted to Master Farmers, to more broadly based groups open to all. The fact that some Farmers Clubs now organize rotating work groups also creates the opportunity for farmers to observe and use a wider array of techniques. However, there is some indication that the Farmers Clubs are still dominated by a rather small group of Master Farmers, who also hold other key public positions. This situation can result in either a real or a perceived control of the flow of some types of agricultural information by certain individuals, who thus take on the role of "gatekeepers." For example, at the end of an interview with someone who was cited as being *mhare/changamuka* (z25), there was an unsolicited comment to the effect that when this farmer wants to teach people something new, the chairman of the Farmers Club (i.e., z201, the most oft-named *mhare/changamuka* farmer) blocks it until he himself is satisfied that the new technique is interesting.

Farmers identified as *mhare/changamuka* are seen as clever and hardworking. The fact that they have access to adequate implements, and are therefore likely to be relatively wealthy, is also recognized as a key to their success. Some of the most oft-named individuals were able to report examples of experimentation, and, indeed, some claimed to have been responsible for significant changes in farming practice. For example, one said he developed a method of ridge making by reverse plowing (z201), and another claimed to have extended the practice of dry seeding to maize (z235). Nevertheless, the trait of "innovativeness" seemed to be of relatively little importance in the public perception of such farmers' success.

In Pamdu in Brong Ahafo, the individuals most commonly named as either successful, dynamic, or those to whom the respondents would go in case of crop disease or problems are described in Table 7.9. As in Chivi, the responses in Pamdu were dominated by a relatively small number of individuals, and the two most oft-named individuals, g202 and g207, combine the same characteristics as their Zimbabwean counterparts: large-scale production, multiple public roles, and technical training in modern agricultural production. At the Awisa site, the most commonly identified local source of agricultural information was a man who is employed by the national Crops Research Institute and who is responsible for establishing and monitoring crop trials. He also sells agrochemicals as a private business activity and advises as to their use.

The two sites in Brong Ahafo present similar pictures in terms of the flow of agricultural information among farmers. In Pamdu, with its very active cadre of young men producing tomatoes as their major economic activity, there were few reported barriers to information move-

ment. Indeed, such comments were common: "We are all tomato growers; we visit each others' farms and take samples of interesting things"
(g204) and, "All tomatoes are grown in the same place; we can see the
other farms; we learn from each other" (g246). There are also indications that the vegetable producers' associations that have been active in
the village for several years have served as a channel for technical and
marketing information. As indicated earlier, some young men report
having learned to grow tomatoes through the weekly discussion meetings of these groups. As in Chivi, these groups may act to circumvent
the problem of secrecy and hearsay by providing a legitimate, public forum for information exchange.

Nevertheless, there is little question that the very competitive nature
of tomato production in both Pamdu and Awisa is associated with certain barriers to the free flow of agricultural information. At both sites, it
was clearly considered improper for *certain* people to "visit" a farm
where they had no immediate business. Straying from the path or road-
side to "inspect" somebody else's farm can itself be seen as malicious.
Thus, the story of one woman, the migrant wife of a schoolteacher who is
himself a large-scale farmer and tractor owner, is not atypical:

> Where I farmed last year there were people who did not go through all
> the processes needed for successful tomato production but their toma-
> toes did well, while I went through all the proper processes and mine
> did not do well. Because of this *juju* [magic] I had to abandon the plot I
> was using. (g226)

It is also interesting to note that the man who is said to have revolu-
tionized tomato production in Pamdu by introducing the Power variety,
using an irrigation pump, and forming a growers' association was also reported to have given up vegetable production and actually to have
moved out of the village after his tomato crops were repeatedly destroyed through *juju*. Clearly, there are very definite barriers to the free
flow of information within Pamdu, and these barriers may become particularly restrictive for strangers, women, and those who are viewed as
too successful. Thus, in the case of Brong Ahafo, the competitive and
commercial nature of tomato production can either facilitate or constrain access to information, depending on who you are.

Our study did not set out to investigate the role of formal groups in
farmers' experimentation or the process of information transfer (see
Bebbington, Merrill-Sands, and Farrington [1994] for a recent review of
this topic). However, the findings from Chivi suggest that in some situations formal groups may play a significant role in the brokerage of agricultural information. What is perhaps most significant is that these
groups do this in a way that is perceived to be sufficiently distinct from

140

Table 7.9. Characteristics of Named Individuals, Brong Ahafo

	Question		
Individual	*"Who is innovative, dynamic, tries new things?"*	*"Who would you go to if you had a problem with tomatoes?"*	Description
g202	5	3	Fifteen years education, former teacher, district assemblyman, Chief Farmer, chairman of PTGA, owns tractor and irrigation pump, sells agrochemicals
g207	1	4	From Sunyani, ten years school plus two years college, joined Dept. of Agriculture in 1970 and rose to level of agricultural supervisor, transferred to Pamdu in 1989 with Global 2000 program, retrenched in 1992, decided to stay in Pamdu and farm; he is also the adult literacy teacher
g209	2	0	Executive member of PTGA, large maize and vegetable producer, wife of g202
g228	1	1	Longtime migrant, ex–Chief Farmer, treasurer of PVPA, large-scale cowpea and vegetable production
g211	1	0	Retired teacher; cited by g202 as an innovator; no additional information
g229	1	0	Treasurer of PTGA, owns irrigation pump, brother of g202, cited by g202 as large grower and innovator
g231	1	0	Cited by g202 as large grower; no additional information
g237	1	0	Teacher, financial secretary of PVPA, not presently farming
g244	1	0	no information available
g246	1	0	Large-scale farmer, owns two tractors and maize sheller, son is tractor

(*continues*)

Table 7.9. *(continued)*

	Question		
Individual	*"Who is innovative, dynamic, tries new things?"*	*"Who would you go to if you had a problem with tomatoes?"*	*Description*
			driver, related to village chief, member of two farming associations
g203	0	1	First went to Nigeria in 1980 and settled again in Pamdu in 1984, chairman of PVPA, former seed producer for Global 2000 program, reported by g237 as formerly the best tomato farmer, worked at the University of Science and Technology/Kumasi and had "the technique," changed how tomatoes were grown, introduced Power variety, used an irrigation pump, but someone cast a juju on his farm and the plants dried up overnight; this happened several times, and finally he left Pamdu in early 1993
g208	0	1	Large eggplant grower, owns kiosk in Pamdu town
g222	0	1	Nine years education, has grown commercial tomatoes for fifteen years, controls large area on land near river, active experimenter
Total	15	11	

"local networking" in order to get around the problems of gossip and suspicion apparently inherent in (at least some) local networks. Formal groups specifically organized for a particular purpose, such as cooperatives, are characterized, in part, by formalized rules, restrictions, and relations; they are therefore often more predictable than informal groups. Paine (1970) argues that for members, formal groups are often easier to deal with precisely because the rules are known. In contrast to formal groups and networks, informal ones often seem to be more open and accessible. However, this appearance of openness can be deceptive as it may simply reflect the fact that the criteria for membership are not as well known. It is also clear that the formal and the informal cannot be so readily separated, especially in small and closely knit communities (Bowen 1993). Bowen also draws attention to the possibility of innovators using channels such as formal groups for disseminating their new ideas, but indicates that there is a potentially significant social cost to the innovator if people do not agree with the idea or think that it is too absurd. The importance of the possibility of "losing face" is also highlighted by Longley and Richards (1993).

With respect to the formal groups we encountered during our fieldwork, membership in the St. Joseph's Group in Kangare was limited, based on clear qualifications, and continually reassessed and enforced (in part, based on a member's willingness to follow a prescribed set of management practices for the dairy cows that were distributed through the group). In such a case, the formal group is used to discuss production problems and exchange information, as well as to enforce certain behavior. This can have both positive and negative effects on the movement and availability of information, and there is clearly potential for gatekeepers to gain some degree of control of information movement. In contrast, membership in the vegetable production associations in Pamdu appeared to be somewhat more open, but the members of the various groups were described as being mostly relatives of the groups' founders. Finally, in Chivi, the Farmers Clubs are becoming so open that the membership fee is no longer seen as a prerequisite for participation.

It is also interesting to note that the farmers in Boma were reported by the local extension officer to have refused to form a local cooperative that might have facilitated access to agricultural inputs and extension services, while an attempt to set up a group of tomato producers in Suminakese was reported to have failed.

FARMERS WHO EXPERIMENT:
EVIDENCE FROM EAST ANGLIA

Moving again to the United Kingdom, the Ideas Competition database provides a measure of the breadth of participation in the contest, which can be used to indicate, at least in a rough way, the distribution of farmers' experimentation or innovative behavior with regard to farm machinery. In fact, participation in the contest appears to be broad based, with 80 percent and 69 percent of entrants in 1948–1994 and 1985–1994, respectively, entering only once. Over both periods, 96 percent of individuals entered fewer than four times. One exceptional individual entered the contest twelve times between 1971 and 1993, while another has entered ten times since 1966 (Knight 1995).

In addressing the question of factors motivating farmers' experimentation, Knight hypothesized that the scale of the farming enterprise was related to the level of interest in experimentation. Specifically, he suggests that smaller farmers are, almost by necessity, more interested in experimentation and innovation, and he illustrates this point accordingly:

> Smaller farmer: "It is the classic thing isn't it, you can buy a new piece of kit [i.e., machinery] one day and out come the gas bottles [for welding] the next! There's always something that needs changing or adapting when you start using things on your own place."

> Larger farmer: "I spend money on a machine and I want it to work, I don't have time to mess around, especially at harvest or rush times." (Ibid.:30)

It is also interesting to note that since 1971, over half of the entries in the contest were from farm employees as opposed to independent farmers (ibid.:24). From this we can conclude that engagement in an agricultural task, whether or not this is accompanied by any significant level of control or management, can in some situations provide sufficient stimulus for experimentation. Thus, if one is not willing to accept the suggestion made earlier that there has been a significant breakdown in the gender roles in African agriculture, then the U.K. data provide another explanation for the observation that there were relatively few differences in terms of the topics and characteristics of experiments reported by men and women.

Carr (1996) provides some interesting insights regarding the spread and larger significance of farmers' innovations in the United Kingdom. Eighteen individuals were asked about the twenty-five innovations they entered in the Ideas Competition between 1988 and 1994. In general, they reported that with the passage of time following the competition they are less likely to continue to use their own innovations, while use by their neighbors, friends, and family tends to increase. In other words, they have already started to abandon an innovation before the local process of diffusion has run its course. Thus, Carr provides an indication that farmers' experiments with machinery have impacts locally; and on a larger scale, the contestants subsequently attempted to patent or commercialize six (24 percent) of these entries. Only two were successfully patented or commercialized, although there were claims that ideas had entered commercial channels behind the backs of the contestants. Such claims must be evaluated in the light of the fact that the competition judges and the farmers themselves recognize that there is nothing new in many of the entries.

SUMMARY

We have previously suggested elsewhere that the design of farmer participatory research activities must take into account the characteristics of both farmers' experiments and the local networks through which agricultural information moves (Okali, Sumberg, and Farrington 1994). The results of our field research presented in this chapter bear on three elements of the discussion of farmers' experiments and local information networks: (1) the characteristics of individuals who experiment; (2) the characteristics of individuals seen to be at the center of the local information network; and (3) the relationship between the local networks and the wider world.

We found no strong relationships between the socioeconomic characteristics we assessed and either the propensity to experiment or the characteristics of the experiments that were reported. Similarly, we found little evidence to support the notion that there is a subpopulation of research-minded farmers who are widely recognized or cited as being central to the generation or movement of information about farming. Although it is certainly true that some individuals are seen as having "special interests" or in-depth knowledge about particular topics (i.e., the local subject matter specialists), this recognition is not apparently associated with the notion of experimentation or innovativeness.

Rather, the people most frequently named by other farmers, are either the larger, male producers who also tend to hold other public positions, or individuals with some formal technical training and a connection to the formal research and/or extension services. And in many cases, frequently named individuals combine both of these sets of characteristics. With a combination of adequate inputs, technical know-how, and valuable institutional links, these individuals have, and are widely seen to have, the keys to successful farming in both a difficult environment (Chivi) and a competitive, market-oriented context (Brong Ahafo).

It is important to note that these are, by and large, the same people with whom the formal research and extension services have worked for many years. They can, after all, carry the economic and social risk of involvement with outside institutions. At the same time, they are generally respected by their neighbors, which is usually presumed to be a prerequisite for a successful demonstration or extension activity. Indeed, much previous research indicates that such individuals play an important role in the dynamic of local agricultural change: They are, in effect, the "opinion leaders" identified and characterized by Rogers (1983:281). They are also the individuals whom some more recent development and participatory research and extension programs specifically try to avoid, as working with such elite groups is seen to be in conflict with equity objectives and the goal of targeting more vulnerable groups.

The notion of multiple and clearly bounded networks that play a role in the movement of agricultural information was not borne out by the research. In fact, the research highlights the close integration of local and nonlocal networks and the fact that a relatively small number of individual farmers seem to provide the key points of reference for most farmers within a particular site. It is, therefore, essential that those interested in farmer participatory research or more flexible approaches to extension determine the actual costs and benefits associated with affirmative steps to extend the range of participants (as opposed to potential beneficiaries). It is also important to recognize that many individuals and communities throughout Africa are now quite directly linked—through family, the government technical service, development programs, local and international radio, and so on—with a whole series of national and international agricultural networks.

While some of the sites visited were more isolated than others, overall, many individuals clearly have access to a relatively wide variety of information sources. These sources include those that might conveniently be referred to as "local," including family, neighbors, the subject matter specialist and the technical agent described above, and projects and extension personnel. Local sources of information also include shops and private suppliers of services, such as veterinarians. These local sources are not always distinct, as many projects utilize government recommendations and personnel, many villagers have participated in project and extension activities, and many suppliers of agricultural inputs and services have close links with formal research and extension and development projects.

At the other end of the spectrum are what might be referred to as "international" sources of information. At Kangare, for example, one individual had used a direct link with a U.S.-based nongovernmental development organization and the local livestock extension office in Rongo to successfully mobilize outside resources that allowed the expansion of dairy production in the village. These links between local rural settings and the outside world are being multiplied and extended from both directions. So, while twenty years of expanding activity by NGOs have dotted projects, local offices, and personnel throughout rural Africa, the migration of Africans both within the continent and beyond provides many families and communities with ever-expanding national and international networks.

8

Farmers, Research, and Extension

In this chapter, we briefly review the major findings and conclusions from the analyses presented in Chapters 6 and 7. These findings are then discussed in relation to, first, the various hypotheses about farmers' experimentation that emerged from the literature review presented in Chapter 3 and, second, the claim of potential synergy from bringing together formal agricultural research and farmers' experiments. We then move on to examine the implications of these findings and conclusions in terms of the participation of farmers in formal agricultural research and extension.

STUDY CONCERNS

Before we proceed, it is important to highlight and briefly address some questions relating to the concepts and definitions used in our study and other questions relating directly to our fieldwork.

The choice of a definition of "experiment" or "experimentation" raises a clear dilemma: Defining "experiment" in the sense of "to try" can be seen as either far too broad (i.e., encompassing many of the things that a farmer does anyway) or too narrow (i.e., not taking into account the *art de la localité* aspects of farming). We tried to get around this problem by extracting from the farmers' narratives situations that corresponded with our definition of experiments (basically, the presence of treatments plus observation). These examples of experiments portray fairly bounded acts, which, although clearly embedded in the larger performance of farming, can nevertheless be described and discussed by the farmers as individual acts.

In the event, almost all of our examples focused on experiments of a very technical nature, and they therefore provide only a limited view of social or institutional experimentation. This bias reflects both our definitions and our methods, and we appreciate that social and institutional

147

experimentation may be both different and potentially more interesting than the examples we identified. However, we are convinced that social and institutional experimentation is very complex and possibly requires skills in management and organization that are unlikely to be widespread within the communities we observed. Two of the examples involved cooperation among individuals in order to address marketing problems: At the time of our fieldwork, one had failed and the outcome of the other was far from certain.

It is clear that agricultural research in the broadest sense encompasses many technical, theoretical, and intellectual strands that are not reflected in our examples of farmers' experiments and that may be beyond the reach of most farmers. Nevertheless, the conditions under which many agricultural researchers and research institutions in Africa function, and the now widely accepted international division of labor within agricultural research, mean that there are many researchers in Africa who are fully occupied by activities that should rightly be seen as adaptive and applied problem solving. It is also important to remember that even in a context such as East Anglia where the formal research system is both better developed and better funded, adaptive and applied research in the form of variety, pesticide rate, and date of planting trials remain, and are likely to always remain, critical research activities.

We are convinced that more intensive work at each of our sites would have allowed deeper and richer analysis. It would have been particularly useful, for example, to explore further the relationships between site-related variables and levels of farmers' experimentation, as well as to further observe the relationship between gender and levels of experimentation. We are also sensitive to the fact that in those cases where we worked through an interpreter, we were on difficult ground in exploring an area as potentially ephemeral as local experimentation.

As is the case with all field research, we were additionally faced with the problem of working with subjective experience. Because of previous criticism of their practices, local people may tell outsiders what they think they want to hear (van Beek 1993). This is certainly not a new concern, as evidenced by the fact that in the 1950s, Eisenstadt (1955) was exploring the normative content of communications. It is true, however, that research relating to agriculture and rural development is now a more self-conscious process (see Long and Long [1992] for a number of articles on this theme). An additional concern was raised in the form of the strong "project mentality" that was evidenced among respondents at some of our sites. These are certainly important factors that characterize the context in which our work took place.

Finally, and perhaps most important, our study sites may not be marginal or isolated enough to have maintained any of their "local research traditions." There is no doubt that we could have found more isolated or more precarious sites in which to work. However, our conclusions are based on a range of conditions that we would argue represent to a reasonable degree those under which many farmers in sub-Saharan Africa produce crops. It is difficult to contest the notion that in some more isolated place, one may identify farmers' experiments with an entirely different character, and there may be some compelling reasons for studying these. Nevertheless, since a large proportion of farmers throughout Africa already have some contact with research and development organizations, use a variety of modern agricultural inputs, and market their crops, it is not evident that such a study would be particularly relevant to our concern with improving the efficiency of agricultural technology generation and transfer.

PRINCIPAL CONCLUSIONS ABOUT FARMERS' EXPERIMENTATION

The evidence presented in Chapter 6 indicates that many farmers, indeed the majority of farmers, experiment as an integral part of their farming activities. Many of these experiments are focused on a limited range of standard agronomic topics, are proactive and planned, have some provision for controls and/or replication, are based on copying what has been observed/suggested or on the experimenters' own ideas, and result in minor modifications to farmers' existing practices or in changes that are already known to, and used by, other local farmers.

From this evidence, we conclude that those farmers' experiments we were able to document, given the time at our disposal and the methods used, have much in common with the agronomic experiments undertaken as part of formal applied and adaptive agricultural research. Both farmers' experiments and much formal experimentation aim to develop practical solutions to immediate problems or to seek small gains within the context of proven production methods and systems. Both are largely empirical and iterative, combining experience, observation (both methodical and opportunistic), intuition, persistence, skill and luck. Neither is concerned primarily, or even to any significant degree, with either the elaboration and testing of theory or the identification of particular physical, biological, or physiological mechanisms that underlay observed results.

Through the examples of farmers' experiments, we have examined one process, or rather one step in a process, by which local knowledge is created. The results of this analysis lend little support to the idea that since all knowledge is socially constructed, there will *necessarily* be major differences in and discontinuities between the way in which so-called formal and local knowledge is created. This conclusion suggests that at least in the case of crop-based agriculture, the subject matter itself may be of much more importance than are the social and cultural contexts in determining the participants, focus, processes, and outcomes of knowledge creation through experimentation.

The field data indicate that although the propensity to experiment is not related in any systematic way to age or education level of farmers, there appears to be some association with gender. There is also some indication that individuals who are poor or otherwise marginal from a socioeconomic perspective may be less able to follow through with experiments because of their lack of control over key productive resources. There is, however, no consistent association between the socioeconomic characteristics of individuals and the kinds of experiments they do.

We have identified a number of site-related factors that may be associated with the level and focus of farmers' experimentation. These factors include the presence or recognition of a particularly difficult problem; level of commitment to farming; openness of the landscape; crop propagation method; life cycle of crops; presence of promotional activities; and the level of commercialization. The interactions among these factors are likely to be multiple and complex.

Data generated both from our interviews and the naming exercises lead us to conclude that there is no well-defined subgroup of farmers whose members are locally recognized as particularly engaged in or skilled at experimentation, although there may be individuals recognized as having particular information, knowledge, or skills. Thus, the image of research-minded farmers as persons who are significantly different from the farming population at large appears to be unfounded. Farmers clearly recognize different styles and scales of farming, and they associate "success" with large-scale and technical advancement as opposed to a particular commitment to experimentation. We observed nothing that would lead us to conclude that the term "informal research and development *system*" has any descriptive or analytical value in relation to farmers' experiments or their interaction with formal research.

HYPOTHESES ABOUT FARMERS' EXPERIMENTS

The review of the literature on farmers' experiments presented in Chapter 3 generated a list of hypotheses, some of which are contradictory and only a few of which were derived from, or are supported by, field studies. We propose now to return to these hypotheses and examine them briefly in the light of our own research findings.

Hypothesis 1: Farmers engage in activities that can be labeled "experiments." As a general statement, this is supported by a substantial body of both systematic and casual observation. Our research in Africa and the studies of farmers' experiments in East Anglia add to the evidence that many farmers actively seek and test potential solutions to practical problems through experiments.

Hypothesis 2.1: Farmers' experiments concern both technical and nontechnical issues. The vast majority of the examples derived from this study focused on technical and, more specifically, agronomic concerns. Some of this bias toward examples of technical experiments reflects the research methods used. Nevertheless, there is also a certain logic in the predominance of technical experiments in that experiments concerned with, for example, marketing, labor arrangements, or other institutional issues are necessarily more complex and may require a level of group action and coordination; thus they carry increased financial and social risk.

Hypothesis 2.2: Farmers' experiments share some characteristics with formal research. Our findings are in agreement with this hypothesis put forward by Rhoades (1989), Scoones and Thompson (1994a), and others. Indeed, we go further and suggest that they share *many* characteristics with adaptive and applied agronomic experiments carried out within formal research: Given the present context of farming in much of Africa—which includes an array of crop varieties and other inputs supplied through government, the private sector, and development organizations, high levels of commoditization and commercialization, increasing levels of education, widespread exposure to research and extension activities, and so forth—one could hardly expect otherwise. Farmers use experiments to help inform concrete decisions that have very real financial and livelihood implications. Compared to researchers, farmers may take a broader or more synthetic view when evaluating technologies, but they still rely essentially on small-scale comparisons that are constructed around a limited number of variables. The fact that their experiments share many characteristics with formal experiments is a reflection of common interests and some basic principles of trying and learning, such as risk minimization, direct comparison, and replication.

The image of an unpolluted, local research tradition does not reflect the reality of many areas in sub-Saharan Africa today. The suggestion that local research traditions have degraded implies that farmers are largely helpless in the face of "formal" knowledge and "Western" research traditions. Although we are aware that these traditions represent powerful interests, we, along with others, have shown that farmers draw together ideas and inputs from many different sources. It is indeed ironic that the image of the helpless or disempowered farmer arises in the context of a discussion of the failure of formal research to make an impact on the majority of farmers in Africa. At the same time, the image of helpless farmers is in stark contrast to the assumed value of farmers' experimentation, an assumption that is itself dependent upon a view of farmers as active agents.

Hypothesis 2.3: Farmers' experiments are socially and culturally embedded; they are based on fundamentally different frameworks compared to formal research. All knowledge and the processes through which knowledge is created and modified are culturally and socially embedded. We are also aware of the warning by Cornwall, Gujit, and Welbourne (1994:100) of the danger of drawing erroneous parallels between formal experimentation and farmers' experimentation. However, this hypothesis has significant implications for any analysis of the potential for synergy between formal research and farmers' experiments. It may well be true that farmers and researchers use different frameworks to interpret their experiments; and, if pushed to do so, they may articulate different explanations of their results, explanations that focus on different mechanisms and relationships. Nevertheless, as argued above, the experiments of adaptive and applied research, on the one hand, and of farmers, on the other, are not about underlying mechanisms and explanations; rather, they are about arriving at practical answers. These conclusions suggest that no matter what differences there are in the deeper origin of, or framework behind, any two experiments, the lessons and implications for the practice of farming in a particular context may be the same. And it is these similarities, rather than the more deeply hidden differences, that will be crucial for assessing potential synergy.

Hypothesis 2.4: Farmers' experiments are more than simply adaptive. Our research indicates that farmers' experiments are *largely* adaptive in that they seek to solve site-specific problems. The majority of farmers' experiments result in relatively minor modifications to existing production methods and systems.

Hypothesis 2.5: Farmers' experiments are not forward-looking. Adaptive research is not, almost by definition, forward-looking, so our research results appear to support the suggestion of Biggs and Clay (1981) that experiments cannot help farmers to "anticipate opportunities." However, although the hypothesis is supported by our data, it contributes little to the advancement of the discussion.

Hypothesis 2.6: Farmers' experiments are haphazard, trial-and-error, and inefficient events and have methodological limitations. Our data indicate that rather than haphazard or reactive, farmers' experiments are largely planned and purposeful undertakings. Although they are, almost by definition, trial-and-error, undertakings, so too is the very nature of even the most formal adaptive and applied research.

We have argued that farmers' experiments are effective in generating site-specific information that, when combined with information from other sources, plays an important role in informing practical farming decisions. We are not suggesting that these experiments might not be improved through, for example, different plot layouts, more replication, and quantification; but, it is not altogether evident that the costs of these changes would be justified, particularly in the eyes of the farmers themselves.

Hypothesis 3.1: Among sites, socioeconomic and agroecological factors either increase (+) or decrease (–) the propensity to experiment: environmental degradation (+); diversification (+); isolation (+/–); access to research and information (–); influence of commodity-oriented research (–); stability (–). Our study provides no evidence with which to address directly the question of the impact of environmental degradation on farmers' experimentation. In a related vein, however, the interviews from Chivi highlighted the association between uncertain rainfall and the farmers' perceived need to experiment. This finding is in line with that of Amanor (1994), who found that degradation at the forest margin essentially forced farmers to be innovative.

The examples of experimentation we identified do not help us understand the relationship between the degree of diversification and the level of experimentation, although the actual process of diversification as a response to changing constraints, needs, and opportunities might well be associated with increased levels of experimentation. One of the key dimensions of diversification in rural livelihoods is a move away from agriculture and other natural resource–based activities, which, we suggest, in itself has important implications for experimentation.

Our study lends little support to the hypothesis that greater contact with formal research and extension, or commodity-oriented research, inhibits local experimentation. Farmers in Chivi, with considerable contact with research and extension programs, appear to experiment actively. A more robust test of this proposition is possible using the U.K. data. The studies from East Anglia show clearly that even in a context where the results of formal research are pervasive and where farmers have access to an abundance of information through numerous channels, they continue to carry out small experiments.

Hypothesis 3.2: Within sites, socioeconomic and personal factors affect the propensity to experiment: poverty (+/–); degree of "research-mindedness" (+). We have suggested that poverty may well increase the need and the desire to experiment, but it can also make it difficult for farmers to complete their experiments. The uncertainty that accompanies limited control over productive resources such as land, labor, or draft animals may, for example, inhibit replication over time. One factor that is perhaps of greater importance than poverty in terms of the propensity to experiment, however, is the degree of commitment to farming. Other things being equal, we conclude from this study that individuals whose livelihoods depend on farming or who farm as a chosen profession are more likely to experiment. At the same time, we found little evidence to support the notion of research-minded farmers as a group distinguishable from the general farming population.

Hypothesis 4: Farmers' experiments need to be, and can be, strengthened or supported. The farmers' experiments considered in this study are not aimed at or designed to arrive at a "best practice" to be applied over an entire village or farm. Rather, they are an essential part of the process by which ideas and technologies—be they new, old, or slightly modified—are tried, tested, rejected, modified, and perhaps slowly integrated into an evolving farming practice. In this sense the value of any particular experiment in any particular year is small, while the value of the accumulated experience is very great. It was also noted that many experiments are initiated and subsequently abandoned as poor rains, lack of draft animals, sickness—the elements of hazard—take their toll.

The need to "strengthen" farmers' experiments is frequently interpreted as a need to formalize them by including, for example, replications, standardized plot size, and quantification of yield. These changes are likely to increase the cost of experimentation and the risk of loss due to the elements of hazard.

Hypothesis 5: Strengthening farmers' experiments empowers farmers. This hypothesis is based on the notion that strengthening farmers' experiments will allow a shift in the power relations between farmers and formal agricultural researchers and their institutions. It may be true that if farmers were more familiar with the jargon, methods, and bureaucratic trappings associated with formal research, then they would be less likely to be intimidated and better able to hold their own in discussions with research and extension personnel. However, we found little evidence that farmers' skills in experimentation constrain their abilities to generate information of value in solving practical problems. In addition, with the added risk associated with increased investment per experiment, this kind of strengthening would appear to be of limited direct value to the majority of farmers.

FARMERS' EXPERIMENTS AND
THE SYNERGY HYPOTHESIS

Hypothesis 6: Closer integration of farmers' experiments and formal research will result in synergistic benefits. A major conclusion of this research is that farmers' experiments have much in common with activities undertaken within formal research institutions under the rubrics of applied and adaptive agricultural research. On reexamination, we also find support for this observation in the literature on farmers' experiments, particularly where actual examples of experimentation are described or analyzed.

In Chapter 3, we distinguished between synergy and complementarity and suggested that the degree of synergy that can be expected is positively related to the degree of difference among the processes to be brought together. If this general proposition is accepted, then we are forced to conclude, in the light of the characteristics of the examples of farmers' experiments, that we should expect little synergistic benefit in bringing farmers' experimentation and formal experimentation into closer contact.

Despite this conclusion, farmers' experiments clearly remain essential for overcoming the all-important, site- and situation-specific dimensions of farming. This constant and widespread experimentation, which takes place within the context of existing production practices, is one important source of the information and knowledge that supports the evolution of agricultural practice and systems.

At the same time, our conclusion concerning the synergy hypothesis should not be confused with the broader conclusion that farmers have no role to play in formal agricultural research.

Hypothesis 7: Farmers' experiments represent an untapped development resource. Much of the literature on farmers' experimentation and indigenous knowledge more generally draws attention to farmers' ideas, techniques, or solutions to problems and suggests that these may have wider application beyond the specific context within which they have been developed or tested. Our data give little support to the suggestion that farmers' experiments represent an untapped source of innovation and technology that over the short term has the potential to change the broader agricultural landscape. Rather, the vast majority of the experiments we identified sought to adapt existing techniques, production systems, and available resources to very local (i.e., plot- or farm-level) conditions. Although the cumulative effect of farmers' experiments may be said to represent broad development potential, there is no indication that farmers are not now experimenting or fully using the results of their experiments.

We suggest that the enthusiasm for the suggestion of broader development potential associated with farmers' experiments results, in part, from a failure to understand the importance of the time dimension. In effect, some of the push for greater participation of farmers in agricultural research, and particularly the new emphasis on farmers' experimentation, is based on the observation that farmers' experiments were, and indeed continue to be, essential to the development and evolution of farming technique and agricultural systems. However, there is a fundamental difference between activities and processes whose cumulative effects result in change over decades or even centuries and those with which a short-term agricultural research program can beneficially interact. In the final analysis, it is the issues of situation- and site-specificity and the extended time horizon that set strict limits on the benefits to be gained from closer integration of farmers' experiments and formal research.

IMPLICATIONS FOR AGRICULTURAL RESEARCH

In rejecting the synergy hypothesis, we are not suggesting either that farmers' experiments are unimportant or that there is nothing that researchers and farmers can gain from closer collaboration. Nor do we conclude that in participating in agricultural research, farmers should be, or should remain, in a subordinate position relative to researchers. However, this study does raise important questions relating to some key elements of the recent discussion of farmers' participation in formal agricultural research.

At one level, applied agricultural research is about improving productivity, quality, or sustainability through the identification, development, testing, and modification of technologies. Whether this research is undertaken by farmers, the government, the private sector, or development organizations, farmers eventually take the research results and determine whether they are of any value in a particular situation.

It is now widely accepted that in order to do their job effectively and efficiently, researchers must "communicate" with farmers. Yet, it must also be recognized that the appropriate level and type of communication and interaction between farmers and researchers must certainly depend on the particular research objective, problem, and context.

In much of the recent discussion of "farmer participatory research," participation is raised from the status of a means to that of an end (if not *the* end). While this might serve certain social development goals and institutional imperatives, it does not necessarily result in the best use of resources available for formal agricultural research; nor does it necessarily provide useful technologies to farmers in a cost-efficient manner. With resources available to research and development severely constrained, the emphasis needs to be squarely on ways of making research more effective; and to reflect this emphasis, a new vocabulary is warranted. We need, for example, to be thinking more in terms of *agricultural research* with farmer participation (of varying types and levels), as opposed to *farmer participatory* research.

We can now return to the distinction introduced in Chapter 3 between research-driven and development-driven farmer participatory research activities. The primary objective of research-driven work is to call on farmers' knowledge, perceptions, skills, and interests in a strategic and flexible way, so as to more efficiently achieve a research objective. The results of our study that are of direct relevance to those engaged in this type of research are twofold: (1) Many farmers are already familiar with the basic premises and principles of experimentation; and (2) in selecting participants, the notion of "research-minded farmers" will be of relatively little value. From a practical point of view, this means that further training of farmers in the basic concepts of experimentation can probably be kept to a minimum as the ideas of side-by-side comparison and replication over time appear to be widely appreciated. In relation to the question of selecting participants, our findings do not lead us to suggest that this should necessarily be at random, or that one might not have specified objectives in selecting particular individuals. For example, differences in communication skills, position in the community, availability, or level of esteem in which certain individuals are held may be useful in order to increase levels of awareness and impact (see, for example, Ashby [1992]). On the other hand, from this work we have seen little indication that selection, either by researchers or by other farmers, of individuals for their level of engagement in experimentation is likely to be worthwhile since the majority of people experiment.

The innovative work within the Rwandan national bean-breeding program, which has been so clearly documented by Sperling and her colleagues at the *Institut des Sciences Agronomiques du Rwanda* (Sperling, Loevinsohn, and Ntabomvura 1993), provides a model of agricultural research with strategic farmer participation. The objective throughout is clear: to make formal research more responsive to the needs and interests of potential users in order to make it more successful and cost effective. In the Rwanda example, there is no question of participation becoming an end in itself, and the participation of farmers did not eliminate the need for subsequent extension and dissemination activities. We suggest that the reason the bean farmers could so readily make a contribution to the program is that the methods used by the researchers were not unfamiliar to them. Thus, the Rwanda example seems to us to point clearly to the complementarity, as opposed to the synergy, of farmers' experiments and formal research.

Our findings have more far-reaching implications for what we have referred to as "development-driven" farmer participatory research. We have provided further evidence that farmers test, evaluate, validate, adapt, and reinvent technologies, that this experience is widespread, and that the tests used have many commonalties with those utilized by formal agronomic research. At the same time, we suggest that these experiments are an integral part of farming and the first step in the adaptation, adoption, and integration of new production techniques. We find nothing to suggest that training of farmers in more formal research methods will have any positive or fundamental impact on their experiments (other than increasing cost and risk), and we therefore conclude that training farmers in the techniques of formal experimentation should not be seen as either essential or desirable.

The need is not to improve the methods that farmers use to experiment, but to increase the supply of "raw material" (i.e., seed of new crop varieties, ideas, etc.) that they can incorporate into their ongoing experimental activities. One important source of that raw material is the formal research system. In this light, it seems clear that the primary objective and the real value of development-driven farmer participatory research will be in getting technology, ideas, varieties, and so forth, into the hands of farmers *sooner* and in a *more flexible form*. The onus for making those varieties, ideas, and technologies work would then lie where it should: with the farmers. A second, but equally important, benefit of this kind of development-driven farmer participatory research is the provision of additional feedback to the formal research system, feedback that may be useful at each of the levels at which decisions about the direction, focus, priorities and forms of research are made.

Seen in this way, development-driven farmer participatory research is first and foremost a particular model of agricultural extension. Projects such as Sustainable Agriculture and Village Extension in Sierra Leone and FarmLink in Egypt (Sumberg 1991; Okali, Sumberg, and Farrington 1994) provide interesting examples of this kind of approach. Some recognition of the value of farmers' experimentation by government research and extension systems is essential if such approaches are to be more widely evaluated.

In terms of the calls for greater participation of farmers in agricultural research, it must be noted that our suggestion of supplying raw material to fuel farmers' experimentation is very different from saying that farmers need to be fully integrated into all stages of the research process, including technology development, or that the division of labor between researchers and farmers must be eliminated. The need for more communication between farmers and researchers, and for strategic participation of farmers within formal research, should now be separated from the suggestion of significant potential synergy and the wide-ranging social development agenda commonly associated with farmer participatory research.

It is critical to remember that the direct outputs of formal research result, in large part, from the specialized knowledge, frameworks, techniques, and tools that are brought to bear on problems and from the access by researchers to a massive body of knowledge and experience. Without these, formal research would have little to offer; at the same time, however, it is clearly unrealistic to propose that farmers should be interested, willing, or able to master all of these *and* continue to farm.

Site selection for the kinds of development-driven activities we are suggesting is critical: Potential participants must be committed to farming and must perceive a problem that they are seriously interested in addressing. As with research-driven farmer participatory research, the notion of "research-minded farmers" is probably of little value in selecting participants.

The dominant version of the international division of labor within agricultural research places national research systems very much at the front line, where technologies either make a contribution or fall away. International policy and the very vocal demands of poverty-oriented development organizations locate that front line in areas that have relatively low agricultural potential. National systems do not have the shelter and comfort offered by the mantle of "strategic" research. Unfortunately but understandably, these institutions are not in a position to respond effectively to demands for low-input, sustainable technology for poor people in complex, diverse, and risk-prone areas. It is, however, unjustifiable to use this fact to force researchers to abandon their disciplines, frameworks, and methods. The influence of the poverty agenda and development organizations, combined with the participatory rural appraisal tidal wave and its associated calls for a "new professionalism," threatens to further restrict any potential contribution of agricultural research.

Finally, there is no question that agricultural research needs to become more effective and more meaningful to its clients, and part of the answer will be greater client orientation through strategic participation of farmers. However, it must remain as research. The current attempts to radically reorient agricultural research to directly serve social development objectives such as participation and empowerment will, in our view, only result in disappointment and long-term loss.

IMPLICATIONS FOR AGRICULTURAL EXTENSION

Agricultural extension in Africa, and throughout much of the developing world, has been criticized, and in many cases rightly so, for being top-down, rigid, and *dirigiste*. Too much emphasis has been placed on detailed packages of recommendations and the delivery of inflexible "messages."

In more recent years, however, there has been some acceptance of the need for more flexibility, such as menus of options, range as opposed to point recommendations, and the unpacking of technical packages. Similarly, there is a growing recognition that movement from an extension recommendation to a change in farming practice usually involves a farmer in a more or less active process of testing and adaptation or reinvention. It is highly ironic, then, that formal research and extension continue to be criticized on the basis of the observation that farmers adopt elements of a technology selectively or that they actively modify and adapt technologies (McCorkle 1989). Is this not both expected and desired?

In a sense, much of the recent discussion of farmer participatory research simply continues to blur the distinctive roles of research and extension, a process that started in earnest with the farming systems research movement. Part of the confusion in the farmer participatory research literature is that many of the activities that are undertaken as development-driven farmer participatory projects consist essentially of microlevel testing and adaptation. Farmers' experiments, on the one hand, and formal adaptive research, on the other, operate at very different levels of specificity. As the formal research process moves past problem definition, their relationship is, and will necessarily remain, essentially sequential, in that farmers' experiments work and rework the outputs of formal research (which themselves may have originated in farmers' experience). We stress again that this should not be interpreted as a call for a reemergence of top-down or one-way models but, rather, as a recognition of the unique conditions within which each and every farmer operates and the unique and valuable microlevel knowledge and skills that farmers use to address these conditions.

We have suggested that the idea of feeding raw material to farmers for their own experimentation is more closely related to the traditional notion of extension than to research. The point is that farmers' site-specific testing of available technology, possibly followed by some adaptation, is a necessary part of the transition from a research-informed extension recommendation to farm practice. Recognition of the pervasiveness and characteristics of farmers' experiments leads to a powerful argument for more open and flexible extension strategies. Such strategies would interface with farmers' experimental activities vis-à-vis the technologies being "promoted." This is what farmers do in any case, and it only makes sense for this important process to be acknowledged, and indeed embraced, by extension.

In other words, while much of the diagnosis and rationale behind the interest in more participation of farmers may be justified, many of the proposed solutions and implementation strategies must be questioned. Strategic participation of farmers is certainly essential for an effective and efficient formal research system. And the need is not for less research or for research that is less "rigorous," but for more and higher-quality research. Equally, there is a real need and opportunity for greater participation on the extension side. High-quality agricultural extension should be defined in terms of its success in fueling farmers' own search for site- and situation-specific solutions and opportunities. This implies a much more open approach, in which technologies coming from formal research (which may themselves be built on ideas from farmers) are viewed as partially cut diamonds that the farmer will polish to fit a particular setting (and may eventually reject, recut, and repolish as that setting changes). In this regard, we are in full agreement with Johnson (1972) when he suggests that "development agencies . . . would be well advised, in spite of the urgency of their goals, to encourage the natural process of innovation by low-risk experimentation, which preserves adaptive diversity and allows proper testing under local conditions" (p. 157).

Such an approach would go a long way toward relieving what is in many ways the very unrealistic and unhelpful pressure that is being applied to national agricultural research systems in Africa. By creating opportunities for *strategic* participation of farmers in agricultural research, and by seeing farmers' experiments as the logical "complement" to formal extension, researchers and research institutions will be better able to make efficient use of limited resources.

However, the more open and flexible approach to extension that we envisage is not necessarily synonymous with the third-generation models described by Rogers (1993), as it does not, in our view, imply or demand that "extension staff need to become 'insiders' rather than 'outsiders'" (p. 16). Nor would NGOs necessarily be better situated than government extension services to develop and implement a more open approach.

Engel (1990) highlights the need for researchers and extension personnel to listen rather than to talk if the agricultural technology system is to move beyond the diffusion of information from research to farmers. However, listening to farmers is still only part of the story as farmers, consumers, merchants, policymakers, and indeed the researchers and extensionists themselves each have a legitimate right to try to affect research priorities and extension recommendations. The real challenge is in balancing these.

At the end of the day, individual farmers will decide the particular suite of agricultural practices they will use, and their own experiments will continue to inform those decisions. Further research leading to a realistic understanding of the role of farmers' experiments should therefore be welcomed by all those involved in the planning and implementation of agricultural research and extension, and in the study of rural development and agrarian change more generally.

Afterword

From time to time over the years that we have studied farmer participatory research and farmers' experimentation, we became concerned that in the end we would be considered naive for taking it at all seriously. Farmers do experiments, and these have significant untapped development potential; farmers' research traditions are being eroded by contact with the modern world; farmers must be involved in all aspects of the technology development process; farmers must drive the research process; there must be no division of labor between farmers and researchers. Should these statements be considered anything other than political and institutional posturing, and if not, are they only of peripheral interest to those involved in the study and practice of agricultural and rural development?

One reason we continue to believe it worthwhile to explore the issue of farmers' experimentation is our sense that it exemplifies an important shift in the terrain on which agricultural and rural development issues are being debated. As such, we thought that this study of farmers' experimentation would highlight important issues and trends that should be of interest beyond the narrow bounds of the debates about agricultural research and extension.

As illustrated in Chapter 4, the fuel for the discussion of farmers' experimentation is drawn from several sources, including purely academic research, more applied research that seeks to bridge the academic and policy domains, project experience, and pure advocacy. At another level, much of its momentum is derived from broader intellectual, political, ideological, and institutional themes and shifts, such as the decreasing role of government and interests in participation and civil society, decentralization, sustainability, local knowledge, and so forth. This is a complex and potent mix that reflects what is potentially a very fundamental shift in the balance of power among the individuals, groups, and institutions involved in planning, teaching, and implementing rural development.

One aspect of this shift in the balance of power that is evident in the debate about participatory research and farmers' experimentation is played out in terms of who contributes and in what form contributions are made. For example, since the mid-1970s, NGOs have come to play an increasingly important role in the design and implementation of rural development activities. It is widely acknowledged that many of the person-

163

nel associated with these organizations—be they recruited locally or internationally—are neither particularly interested nor well placed (because of educational background and/or lack of access to material) to contribute to the larger debates. At the same time, many researchers involved in the more academic side of the disciplines associated with development studies have limited knowledge of, or access to, the experience of practical development work. These two groups share few common frameworks or points of reference, and there is little agreement as to a core body of literature, theory, or principles.

Farmer participatory research and farmers' experimentation have recently enjoyed a certain prominence because this gap between theory and practice, research and action, has been bridged in a very special way. Through several key conferences and their published proceedings, and through the work of a handful of international newsletters and networks, the larger debate has been made accessible to even the most isolated development workers. Simultaneously, the experiences of development projects have been made available to an audience of academics and policymakers who might otherwise have had only limited access. The result has been the creation of a very "positive feedback loop" and a growing, largely self-referencing body of literature that is neither academic nor strictly applied.

We are both intrigued and concerned by this bridging and networking process. It has taken on a life of its own, resulting in a situation where particular contributions, and the topic as a whole, are judged by neither academic nor practical criteria. And, in a sense, they cannot be judged at all; for, as we have already mentioned, there are few common frameworks or bodies of shared understanding. Indeed, it would seem that discussion of, or agreement on, even the most basic vocabulary is limited. We are reminded of an exchange that took place at the 1987 workshop that resulted in the publication of the influential book *Farmer First* (Chambers, Pacey, and Thrupp 1989). A representative from an NGO promoting the use of farmer participatory research made firm and repeated claims that the NGO's work had resulted in increases in the maize yields of small-scale producers of between 200 percent and 400 percent. When this claim was queried, the individual could supply no actual yield data and, indeed, appeared to see no reason why such a request would be made. For our purpose, it is critical to note that this largely undocumented experience has come to rest at the very center of the subsequent farmer participatory research literature.

We suggest that this situation has come about largely because of the critical role played by a small number of intermediary organizations. These organizations have helped create opportunities for the coming together of the academic and the applied, the formal and the informal, and the technical and the social sides of the discussion. They have then edited, grouped, synthesized, and helped publish and disseminate the resulting work. Processes such as these have yielded many of the key texts within the literature on farmer participatory research (and we have made use of many of them in our own work).

What is interesting and important about this process is the role these intermediary organizations play in orientating the discussion and in legitimizing lines and elements of the debate. In effect, they have become central to the dynamic by which ideas and examples become an accepted part of the larger discussion. One key to the centrality of their role is the fact that they seek out, publish, and distribute large quantities of written material. This circulation of written material has become perhaps the most important aspect of international networking. There are, however, generally few guidelines or procedures for the review of potentially published material, and in many cases, the review process involves little input from outside the intermediary organization itself. We suggest that this is in large part responsible for both the rapid growth and the particular characteristics (e.g., little agreement on even basic vocabulary, recycling of a limited number of examples) of much of the farmer participatory research literature we have reviewed for this book.

In Chapter 7, we indicated that there were important similarities between the movement of agricultural information through local networks and gossip. In a sense, the networking activities of these intermediary organizations also have at least two characteristics in common with gossiping: Much of the information is of undetermined quality, and the choice of what information to pass on and in what form can reflect an institutional agenda that may or may not be apparent to other members of the network.

This poses a very real problem for those interested in either directly affecting or studying processes of development. In the case of farmer participatory research, the promotional orientation of much of the literature makes informed implementation decisions almost impossible. At the same time, from a more academic point of view, there is a question as to how far one should engage directly with this literature. As illustrated by our work with farmers' experiments, the particular qualities of the participatory research literature make it difficult for it to be integrated with literature and information from other sources.

In the end, we persevered with our study of farmer participatory research and farmers' experiments despite the fact that as topics of discussion they are largely the products of the dynamic outlined above. We remain convinced that this dynamic has, and will likely continue to have, important impacts on formal agricultural research and extension institutions and activities. It is now an established part of development discourse and development activity, and as such, it is something with which both academics and practitioners must reckon.

Appendix

Table A.1. Characteristics of Respondents by Site (% within site)

Characteristic	Kangare	Chivi	Eastern Region	Brong Ahafo	Overall
Gender					
Male	64	41	60	62	56
Female	36	59	40	38	44
$n =$	14	49	73	53	189
Age					
15–35	21	22	33	42	32
36–55	36	59	36	40	43
56+	43	18	32	19	25
$n =$	14	49	73	53	189
Education (yrs.)					
0–4	27	41	74	40	53
5–8	36	55	25	14	30
9+	36	4	1	46	17
$n =$	11	49	72	52	184
Marital Status					
Never married	0	2	6	12	6
Divorced/widowed	29	32	24	17	24
Married	71	67	71	71	70
$n =$	14	49	72	52	187
Resident spouse					
Yes	90	73	88	89	85
No	10	27	12	11	15
$n =$	10	33	50	37	130
Primary occupation					
Farming	71	88	90	93	89
Other	29	12	10	8	11
$n =$	14	49	73	53	189
Remittances received					
Yes	56	43	39	17	35
No	44	57	61	83	65
$n =$	9	49	72	47	177
Worked outside area					
Yes	88	55	51	47	53
No	13	45	49	53	47
$n =$	8	49	72	51	180

(continues)

Table A.1. (*continued*)

Characteristic	Kangare	Chivi	Eastern Region	Brong Ahafo	Overall
Worked with extension					
Yes	46	35	14	10	20
No	55	65	86	90	80
n =	11	49	72	51	183
Example					
Yes	29	39	73	17	45
No	71	61	27	83	55
n =	14	49	73	53	189

Table A.2. Characteristics of Respondents by Gender (% within gender)

Characteristic	Male	Female	Overall
Age			
15–35	32	31	32
36–55	40	47	43
56+	28	23	25
n =	106	83	189
Education (yrs.)			
0–4	45	63	53
5–8	29	32	30
9+	26	5	17
n =	103	81	184
Marital Status			
Never married	8	4	6
Divorced/widowed	5	49	25
Married	87	47	69
n =	101	81	182
Resident spouse			
Yes	96	59	85
No	4	41	15
n =	91	39	130
Primary occupation			
Farming	90	88	89
Other	10	12	11
n =	106	83	189
Remittances received			
Yes	28	43	35
No	72	57	65
n =	96	81	177
Worked outside area			
Yes	72	28	53
No	28	72	47
n =	102	78	180
Worked with extension			
Yes	28	10	20
No	72	90	80
n =	102	81	183
Example			
Yes	64	43	55
No	36	57	45
n =	106	83	189

References

Abedin, Z., and F. Haque. 1991. Learning from Farmer Innovations and Innovators Workshops: Experiences from Bangladesh. In B. Haverkort, J. van der Kamp, and A. Waters-Bayer, eds., *Joining Farmers' Experiments: Experiences in Participatory Technology Development*. London: Intermediate Technology (IT) Publications.

Aboyade, B. O. 1987. *The Provision of Information for Rural Development*. Ibadan: Fountain Publications.

Amanor, K. S. 1990. Analytical Abstracts on Farmer Participatory Research. Agricultural Administration (Research and Extension) Occasional Paper No. 10. London: Overseas Development Institute (ODI).

Amanor, K. S. 1993a. Farmer Experimentation and Changing Fallow Ecology in the Krobo District of Ghana. In W. de Boef, K. Amanor, and K. Wellard, eds., *Cultivating Knowledge*. London: IT Publications.

Amanor, K. S. 1993b. *Wenchi Farmer Training Project: Social/Environmental Baseline Study*. London: Overseas Development Administration (ODA).

Amanor, K. S. 1994. *The New Frontier: Farmers' Response to Land Degradation, A West African Study*. London: Zed Books.

Anderson, J., ed. 1994. *Agricultural Technology: Policy Issues for the International Community*. Wallingford: CAB International.

Anderson, J. R., R. W. Herdt, and G. M. Scobie. 1988. *Science and Food: The CGIAR and Its Partners*. Washington, D.C.: The World Bank.

Anon. 1991. A Household Welfare Monitoring and Evaluation Survey of South Nyanza District. Unpublished report.

Anthony, K. R. M., and V. C. Uchendu. 1970. *Agricultural Change in Mazabuka District, Zambia*. Stanford: Food Research Institute, Stanford University.

Ashby, J. A. 1992. Identifying Beneficiaries and Participants in Client-Driven On-Farm Research. *Twelfth Annual Farming Systems Symposium*. East Lansing: Michigan State University.

Ashby, J. A., C. A. Quiros, and Y. M. Rivers. 1989. Experiences with Group Techniques in Colombia. In R. Chambers, A. Pacey, and L. A. Thrupp, eds., *Farmer First: Farmer Innovation and Agricultural Research*. London: IT Publications.

Ashcroft, J., and R. Agunga. 1994. Diffusion Theory and Participatory Decision Making. In S. A. White, K. S. Nair, and J. Ascroft, eds., *Participatory Communications: Working for Change and Development*. New Delhi: Sage Publications.

Aumeeruddy, Y. 1995. Phytopractices: Indigenous Horticultural Approaches to Plant Cultivation and Improvement in Tropical Regions. In D. M. Warren, L. J. Slikkerveer, and D. Brokensha, eds., *The Cultural Dimension of Development: Indigenious Knowledge Systems*. London: IT Publications.

Baah, F., Y. Baguna, J. Dijkman, A. Karanja, J. Rua, N. Sutrisno, and H. Waaijenberg. 1994. *Perceptions of Agriculture: A Study of Two Villages in Suhun District, Eastern Region of Ghana*. Wageningen: International Course for Research on Development-Oriented Agriculture (ICRA).

Bagchee, A. 1994. Agricultural Extension in Africa. World Bank Discussion Papers, Africa Technical Department Series No. 231. Washington D.C.: World Bank.

Balderrama, S., T. Fenta, A. Hussein, S. Jackson, M. Midre, C. Scott, C. Vasquez, and J. de Vos. 1987. *Farming Systems Dynamism and Risk in a Low Potential Area: Chivi South, Masvingo Province, Zimbabwe*. Wageningen: ICRA.

Bascom, W. R. 1948. Ponapean Prestige Economy. *Southwestern Journal of Anthropology*, 4:211–222.

Batterbury, S. 1993. Planners or Performers? Indigenous Dryland Farmers in Northern Burkina Faso. Agricultural Administration (Research and Extension) Network Paper 42. London: ODI.

Bebbington, A. J., D. Merrill-Sands, and J. Farrington. 1994. Farmer and Community Organisations in Agricultural Research and Extension: Functions, Impacts, and Questions. Agricultural Administration (Research and Extension) Network Paper 47. London: ODI.

Bentley, J. W. 1994. Facts, Fantasies, and Failures of Farmer Participatory Research. *Agriculture and Human Values*, 11:140–150.

Biggs, S. D. 1980. Informal R&D. *CERES*, 13:23–27.

Biggs, S. D. 1989. Resource-Poor Farmer Participation in Research: A Synthesis of Experiences from Nine Agricultural Research Systems. On-Farm Client-Oriented Research (OFCOR) Comparative Study Paper No. 3. The Hague: International Service for National Agricultural Research (ISNAR).

Biggs, S. D. 1990. A Multiple Source of Innovation Model of Agricultural Research and Technology Promotion. *World Development*, 18(11):1481–1499.

Biggs, S. D. 1994. Farming Systems Research and Rural Poverty: Relationships Between Context and Content. *Agricultural Systems*, 47(2):147–159.

Biggs, S. D., and E. Clay. 1981. Sources of Innovation in Agricultural Technology. *World Development*, 9(4):321–336.

Biggs, S., and J. Farrington. 1991. *Agricultural Research and the Rural Poor: A Review of Social Science Analysis*. Ottawa: International Development Research Center (IDRC).

Boissevain, J., and J. C. Mitchell. 1973. *Network Analysis: Studies in Human Interaction*. The Hague: Mouton & Co.

Boserup, E. 1970. *Women's Role in Economic Development*. London: Allen & Unwin.

Boster, J. S. 1986. Exchange of Varieties and Information Among Aguaruna Manioc Cultivators. *American Anthropologist*, 88:428–436.

Bowen, P. N. 1993. *Longing for Land: Tradition and Change in a Swazi Agricultural Community*. Aldershot: Avebury Publishing.

Box, L. 1989. Virgilio's Theorem: A Method for Adaptive Agricultural Research. In Chambers, Pacey, and Thrupp, eds., *Farmer First*.

Box, T. W. 1967. Range Deterioration in West Texas. *Southwestern Historical Quarterly*, 9:37–45.

Brammer, H. 1980. Some Innovations Don't Wait for Expertise: A Report on Applied Research by Bangladeshi Peasants. *CERES*, 13(2):24–29.

Bratton, M. 1994. Land Redistribution, 1980–1990. In M. Rukuni and C. K. Eicher, eds., *Zimbabwe's Agricultural Revolution*. Harare: University of Zimbabwe Publications.

Brown, D. 1994. Strategies of Social Development: Non-governmental Organisations and the Limitations of the Freirean Approach. Department of Agricultural Extension and Rural Development: The New Bulmershe Papers. Reading: University of Reading.

Bryceson, D. F. 1995a. African Women Hoe Cultivators: Speculative Origins and Current Enigmas. In D. F. Bryceson, ed., *Women Wielding the Hoe: Lessons from Rural Africa for Feminist Theory and Development Practice*. Oxford: Berg Publishers.

Bryceson, D. F. 1995b. Burying the Hoe? In Bryceson, ed., *Women Wielding the Hoe*.

Budelman, A., A. Buchukundi, and F. Mizambwa. 1996. Farmer Opinions on Change and Experimentation in North Sukumaland. In A. Budelman, ed., *In Search of Sustainability: Nutrients, Trees, and Farmer Experimentation in North Sukumaland Agriculture*. Working Paper 16, Tanzania/Netherlands Farming Systems Research Project. Amsterdam: Royal Tropical Institute.

Bunch, R. 1989. Encouraging Farmers' Experiments. In Chambers, Pacey, and Thrupp, eds., *Farmer First*.

Bunch, R. 1991. People-Centred Agricultural Improvement. In Haverkort, van der Kamp, and Waters-Bayer, eds., *Joining Farmers' Experiments*.

Busch, L., R. J. Bingen. 1994. Restructuring Agricultural Research: Some Lessons from Experience. Briefing Paper 13. The Hague: ISNAR.

Byerlee, D. 1994. Technology Transfer Systems for Improved Crop Management: Lessons for the Future. In J. Anderson, ed., *Agricultural Technology: Policy Issues for the International Community*. Wallingford: CAB International.

Carr, N. 1996. The Invention to Diffusion Process: The Significance of Farmers' Innovations in the Development of Farm Machinery. BSc diss., School of Development Studies, University of East Anglia, Norwich.

Central Statistics Office (CSO) (1991). *1982 Population Census: Compilations by Province, District Councils, and Enumeration Areas*. Harare: CSO.

Chambers, R. 1989. Reversals, Institutions, and Changes. In Chambers, Pacey, and Thrupp, eds., *Farmer First*.

Chambers, R. 1993. *Challenging the Professions: Frontiers for Rural Development*. London: IT Publications.

Chambers, R., and B. P. Ghildyal. 1985. Agricultural Research for Resource-Poor Farmers: The Farmer-First-and-Last Model. *Agricultural Administration and Extension*, 20:1–30.

Chambers, R., A. Pacey, and L. A. Thrupp, eds. 1989. *Farmer First: Farmer Innovation and Agricultural Research*. London: IT Publications.

Chasi, M., and Z. Shamudzarira. 1992. Agro-ecologies of the Small-Scale Farming Areas of Zimbabwe. In E. E. Whingwiri, M. Rukuni, K. Mashingaidze, and C. M. Matanyaire, eds., *Small-Scale Agriculture in Zimbabwe: Book One, Farming Systems, Policy, and Infrastructure Development*. Harare: Rockwood Publishers.

Chilver, A. S., and R. Suherman. 1994. Patterns, Processes and Impacts of Technology Diffusion: The Case of True Potato Seed (TPS) in Indonesia. *Journal of the Asian Farming Systems Association*, 2:285–304.

Chowdhury, M. K., and E. H. Gilbert. 1996. Reforming Agricultural Extension in Bangledesh: Blending Greater Participation and Sustainability with Institutional Strengthening. Agricultural Research and Extension Network Paper No. 61. London: ODI.

Cleaver, K. M., and W. G. Donovan. 1995. Agriculture, Poverty, and Policy Reform in Sub-Saharan Africa. World Bank Discussion Papers, Africa Technical Department Series No. 280. Washington, D.C.: World Bank.

Cohen, A. P. 1993. Segmentary Knowledge: A Whalsay Sketch. In M. Hobart, ed., *An Anthropological Critique of Development*. London: Routledge.

Conklin, H. C. 1957. Hanunoo Agriculture in the Philippines. Forestry Department Paper 12. Rome: Food and Agriculture Organization (FAO).

Connell, J. 1991. Farmers' Experiments with a New Crop. In Haverkort, van der Kamp, and Waters-Bayer, eds., *Joining Farmers' Experiments*.

Corbett, J. 1994. *Livelihoods, Food Security and Nutrition in a Drought Prone Part of Zimbabwe*. Oxford: Centre for the Study of African Economies, University of Oxford.

Cornwall, A., I. Gujit, and A. Welbourn. 1994. Acknowledging Process: Methodological Challenges for Agricultural Research and Extension. In I. Scoones and J. Thompson, eds., *Beyond Farmer First*. London: IT Publications.

Coughenour, C. M., and S. M. Nazhat. 1985. *Recent Change in Villages and Rainfed Agriculture in Northern Central Kordofan: Communication Process and Constraints*. Lexington: Department of Sociology, University of Kentucky.

de Boef, W., K. Amanor, and K. Wellard. 1993. Introduction. In de Boef, Amanor, and Wellard, eds., *Cultivating Knowledge*.

de Schlippe, P. 1956. *Shifting Cultivation in Africa: The Zande System of Agriculture*. London: Routledge & Kegan Paul.

Delgado, C. J., J. W. Mellor, and M. J. Blackie. 1987. Strategic Issues in Food Production in Sub-Saharan Africa. In J. W. Mellor, C. L. Delgado, and M. J. Blackie, eds., *Accelerating Food Production in Sub-Saharan Africa*. Baltimore: Johns Hopkins University Press.

Dunkel, F. 1985. Rwanda Local Crop Storage / FSM II. Unpublished report.

Echeverría, R. G., ed. 1990. *Methods for Diagnosing Research System Constraints and Assessing the Impact of Agricultural Research. Volume II: Assessing the Impact of Agricultural Research*. The Hague: ISNAR.

Eicher, C. K., and M. Rukuni. 1994. Zimbabwe's Agricultural Revolution: Lessons for Southern Africa. In Rukuni and Eicher, eds., *Zimbabwe's Agricultural Revolution*.

Eisenstadt, S. N. 1955. Communication Systems and Social Structure: An Exploratory Comparative Study. *Public Opinion Quarterly*, 19:153–167.

Engel, P. 1990. Two Ears, One Mouth: Participatory Extension or Why People Have Two Ears and Only One Mouth. *AT Source*, 18:2–5.

Eponou, T. 1993. Partners in Agricultural Technology: Linking Research and Technology Transfer to Serve Farmers. Research Report. The Hague: ISNAR.

Evans-Pritchard, E. E. (1940). *The Nuer: A Description of the Modes of Livelihood and Political Institutions of a Nilotic People*. Oxford: Oxford University Press.

Everson, R. E., and C. E. Pray. 1991. *Research and Productivity in Asian Agriculture*. Ithaca: Cornell University Press.

Ewell, P. 1990. Links Between On-Farm Research and Extension in Nine Countries. In D. Kaimowitz, ed., *Making the Link: Agricultural Research and Technology Transfer in Developing Countries*. Boulder: Westview Press.

Eyzaguirre, P. 1992. Farmer Knowledge, World Science, and the Organization of Agricultural Research Systems. In J. L. Moock and R. E. Rhoades, eds., *Diversity, Farmer Knowledge, and Sustainability*. Ithaca: Cornell University Press.

Fairhead, J. 1990. Fields of Struggle: Towards a Social History of Farming Knowledge and Practice in a Bwisha Community, Kivu, Zaire. Ph.D. diss., School of African and Oriental Studies, University of London, London.

Farrington, J., and A. Bebbington. 1993. *Reluctant Partners? Non-governmental Organisations, the State, and Sustainable Agricultural Development*. London: Routledge.

Farrington, J., and A. Martin. 1988. Farmer Participation in Agricultural Research: A Review of Concepts and Practices. Agricultural Administration Unit Occasional Paper No. 9. London: ODI.

Farrington, J., and S. B. Mathema. 1991. Managing Agricultural Research for Fragile Environments: Amazon and Himalayan Case Studies. Agricultural Administration Unit Occasional Paper 11. London: ODI.

Ford, S. A., and E. M. Babb. 1989. Farmer Sources and Use of Information. *Agribusiness*, 5(5):465–476.

Franzel, S., L. Hitimana, and E. Akyeampong. 1995. Farmer Participation in On-Station Tree Species Selection for Agroforestry—A Case-Study from Burundi. *Experimental Agriculture*, 31(1):27–38.

Freemen, J. D. 1955. Iban Agriculture: A Report on the Shifting Cultivation of Hill Rice by the Iban of Sarawak. Colonial Research Studies No. 18. London: Her Majesty's Stationery Office.

Gamser, M., and H. Appleton. 1995. Tinker, Tiller, Technical Change: Peoples' Technology and Innovation Off the Farm. In Warren, Slikkerveer, and Brokensha, eds., *The Cultural Dimension of Development*.

Garforth, C. 1982. Reaching the Rural Poor: A Review of Extension Strategies and Methods. In G. E. Jones and M. J. Rolls, eds., *Progress in Rural Extension and Community Development*. Chichester: John Wiley & Sons.

GFA. 1986. Study on the Economic and Social Determinants of Livestock Production in the Communal Areas—Zimbabwe. Hamburg: Geseuschaft für Agrarprojektein Übersee (GFA).

Gilbert, E. 1995. The Meaning of the Maize Revolution in Sub-Saharan Africa: Seeking Guidance from Past Impacts. Agricultural Administration (Research and Extension) Network Paper 55. London: ODI.

Gilbert, E., P. Matlon, and P. Eyzaguirre. 1994. New Perspectives for Vulnerable Institutions: Agricultural Research Systems in the Small Countries of West Africa. Discussion Paper No. 94–17. The Hague: ISNAR.

Gilbert, E., L. Phillips, W. Roberts, M. Sarch, M. Smale, and A. Stroud. 1994. Maize Research Impact in Africa: The Obscured Revolution. Washington, D.C.: Office of Sustainable Development, USAID.

Gilbert, E., J. Posner, and J. Sumberg. 1990. Farming Systems Research Within a Small Research System: A Search for Appropriate Models. *Agricultural Systems*, 33:327–346.

Goldman, A. 1991. Tradition and Change in Postharvest Pest Management in Kenya. *Agriculture and Human Values*, 8(1–2):99–113.

Gubbels, P. 1988. Peasant Farmer Agricultural Self-Development. *(ILEIA) Newsletter*, 4(3):11–14.

Gubbels, P. 1993. Peasant Farmer Organisation in Farmer-First Agricultural Development in West Africa: New Opportunities and Continuing Constraints. Agricultural Administration (Research and Extension) Network Paper 40. London: ODI.

Hall, J. C. 1994. The Transfer of Agricultural Information in Rural Areas: An Anthropological Perspective of Farmers' Networks. Unpublished report. London: ODI.

Hanks, L. M. 1972. *Rice and Man: Agricultural Ecology in Southeast Asia*. Honolulu: University of Hawaii Press.

Haverkort, A. W. 1988. Agricultural Production Potentials: Inherent or the Result of Investments in Technology Development? *Agricultural Administration & Extension*, 30:127–141.

Haverkort, B. 1991. Farmers' Experiments and Participatory Technology Development. In Haverkort, van der Kamp, and Waters-Bayer, eds., *Joining Farmers' Experiments*.

Haverkort, B., J. van der Kamp, and A. Waters-Bayer, eds. 1991. *Joining Farmers' Experiments: Experiences in Participatory Technology Development.* London: IT Publications.

Heinrich, G. M. 1993. Strengthening Farmer Participation Through Groups: Experiences and Lessons from Botswana. OFCOR Discussion Paper 3. The Hague: ISNAR.

Hill, P. 1963. *Migrant Cocoa-Farmers of Southern Ghana.* Cambridge: Cambridge University Press.

Howell, J. 1988. Training and Visit Extension in Practice. ODI Agricultural Administration Unit, Occasional Paper 8. London: ODI.

Howes, M. 1979. The uses of indigenous technical knowledge in development. *IDS Bulletin,* 10(2):12–23.

Hulme, D. 1991. Agricultural Extension Services as Machines: The Impacts of the Training and Visit Approach. In W. Gustafsen and D. Gustafsen, eds., *Agricultural Extension: Worldwide Institutional Evolution and Forces for Change.* New York: Elsevier.

Hutchinson, J., and A. C. Owers. 1980. *Change and Innovation in Norfolk Farming.* Chichester: Packard Publishing Ltd.

Idowu, O., and J. I. Guyer. 1991. Commercialization and the Harvest Work of Women in Ibarapa, Oyo State, Nigeria. Women's Research and Documentation Centre Occasional Paper No. 2. Ibadan: Institute of African Studies, University of Ibadan.

Izikowitz, K. G. 1951. Lamet: *Hill Peasants in French Indochina* (Ethnologiska studier 17). Goteborg, Ethnografoska Museet.

Jaetzold, R., and H. Schmidt. 1982. *Farm Management Handbook of Kenya, Volume II/A.* Nairobi: Ministry of Agriculture.

Jahnke, H. E. 1982. *Livestock Production Systems and Livestock Development in Tropical Africa.* Kiel: Kieler Wissenschaftsverlag Vauk.

Johnson, A. W. 1972. Individuality and Experimentation in Traditional Agriculture. *Human Ecology,* 1(2):149–160.

Kaimowitz, D., M. Snyder, and P. Engel. 1990. A Conceptual Framework for Studying the Links Between Agricultural Research and Technology Transfer in Developing Countries. In Kaimowitz, ed., *Making the Link.*

Kandiyoti, D. 1985. *Women in Rural Production Systems: Problems and Policies.* Paris: United Nations Educational, Scientific, and Cultural Organization.

Knight, M. 1995. *Farmer Participation in Agricultural Research: Machinery Innovations in Norfolk Agriculture.* BSc diss., School of Development Studies, University of East Anglia, Norwich.

Lionberger, H. F. 1960. *Adoption of New Ideas and Practices.* Ames: Iowa State University Press.

Lipton, M. 1985. The Place of Agricultural Research in the Development of Sub-Saharan Africa. Discussion Paper 202. Brighton: Institute of Development Studies (IDS).

Lipton, M. 1989. *New Seeds and Poor People.* London: Unwin Hyman.

Lipton, M. 1994. Agricultural Research Investment Themes and Issues: A View from Social-Science Research. In Anderson, ed., *Agricultural Technology.*

Lipton, M., and S. Maxwell. 1992. The New Poverty Agenda: An Overview. Discussion Paper 306. Brighton: IDS.

Long, N., and A. Long, eds. 1992. Battlefields of Knowledge: The Interlocking of Theory and Practice in Social Science Research and Development. London: Routledge.

Longley, C., and P. Richards. 1993. Selection Strategies of Rice Farmers in Sierra Leone. In de Boef, Amanor, and Wellard, eds., *Cultivating Knowledge.*

Lynam, J. K., and M. J. Blackie. 1994. Building Effective Agricultural Research Capacity: The African Challenge. In Anderson, ed., *Agricultural Technology.*

Lyon, F. 1994. *Farmers' Experimentation in East Anglia: Its Nature and Validity.* MSc diss., School of Development Studies, University of East Anglia, Norwich.

Manners, R. 1956. Tabara: Subcultures of a Tobacco and Mixed Crops Municipality. Stewart, J. H., ed., *The People of Puerto Rico.* Urbana: University of Illinois Press.

Maurya, D. M. 1989. The Innovative Approach of Indian Farmers. In Chambers, Pacey, and Thrupp, eds., *Farmer First.*

McClean, S. P. 1991. The Morley Research Centre. *Journal of the Royal Agricultural Society of England,* 152:159–167.

McCorkle, C. M. 1989. Toward a Knowledge of Local Knowledge and Its Importance for Agricultural RD&E. *Agriculture and Human Values,* 6(3):4–12.

McCorkle, C. M., R. H. Brandstetter, and G. McClure. 1988. A Case Study on Farmer Innovation and Communication in Niger. Communication for Technology Transfer in Africa (CTTA). Washington, D.C.: Academy of Educational Development.

McCorkle, C., and G. McClure. 1995. Farmer Know-How and Communication for Technology Transfer: CTTA in Niger. In Warren, Slikkerveer, and Brokensha, eds., *The Cultural Dimension of Development.*

Mehretu, A. 1994. Social Poverty Profile of Communal Areas. In Rukuni and Eicher, eds., *Zimbabwe's Agricultural Revolution.*

Mellor, J. W. 1988. Agricultural Development Opportunities for the 1990s—The Role of Research. Address presented to the International Centres Week of CGIAR, 4 November 1988, Washington, D.C.

Mellor, J. W., C. L. Delgado, and M. J. Blackie. 1987. Priorities for Accelerating Food Production in Sub-Saharan Africa. In Mellor, Delgado, and Blackie, eds., *Accelerating Food Production.*

Merrill-Sands, D., S. Biggs, R. J. Bingen, P. Ewell, J. McAllister, and S. Poats. 1991. Institutional Considerations in Strengthening On-Farm Client-Oriented Research in National Agricultural Research Systems: Lessons from a Nine-Country Study. *Experimental Agriculture,* 27:343–373.

Merrill-Sands, D., P. Ewell, S. Biggs, R. J. Bingen, J. McAllister, and S. Poats. 1992. Management of Key Institutional Linkages in On-Farm Client-Oriented Research. In Moock and Rhoades, eds., *Diversity, Farmer Knowledge, and Sustainability.*

Merrill-Sands, D., and D. Kaimowitz. 1990. *The Technology Triangle: Linking Farmers, Technology Transfer Agents, and Agricultural Researchers.* The Hague: ISNAR.

Millar, D. 1993. Farmer Experimentation and the Cosmovision Paradigm. In de Boef, Amanor, and Wellard, eds., *Cultivating Knowledge.*

Millar, D. 1994. Experimenting farmers in northern Ghana. In Scoones and Thompson, eds., *Beyond Farmer First.* London: IT Publications.

Ministry of Agriculture. 1991. *Agriculture in Ghana: Facts and Figures.* Accra: Presbyterian Press.

Mitchell, J. C., ed. 1969. *Social Networks in Urban Situations: Analysis of Personal Relationships in Central African Towns.* Manchester: Manchester University Press.

Moock, P. 1976. The Efficiency of Women as Farm Managers: Kenya. *American Journal of Agricultural Economics*, 58(5):831–835.

Mooney, P. R. 1993. Exploiting Local Knowledge: International Policy Implications. In de Boef, Amanor, and Wellard, eds., *Cultivating Knowledge.*

Mundy, P. A., and J. L. Crompton. 1995. Indigenous Communication and Indigenous Knowledge. In Warren, Slikkerveer, and Brokensha, eds., *The Cultural Dimension of Development.*

Murphy, M. C. 1995. *Report on Farming in the Eastern Counties of England, 1993/94.* Cambridge: Department of Land Economy, University of Cambridge.

Murwira, K. (n.d.) NGO and government activities in Chivi District, Masvingo. Unpublished report. Harare: ITDG.

Murwira, K., M. Vela, C. Bungu, and N. Mapepa. 1995. The Experiences of the Chivi Food Security Project: The Project Acting as Facilitator. In V. Scarborough, ed., *Farmer-Led Approaches to Extension.* Agricultural Research & Extension Network Paper 59b. London: ODI.

Mutimba, J. 1994. Agricultural Extension Policy and Practice in Zimbabwe: A Study Report. Unpublished report to ITDG, Harare.

Netting, R. M. 1968. *Hill Farmers of Nigeria: Cultural Ecology of the Kofyar of the Jos Plateau.* Seattle: University of Washington Press.

Norman, D., D. Baker, G. Heinrich, and F. Worman. 1988. Technology development and farmer groups: experiences from Botswana. *Experimental Agriculture*, 24(3):321–331.

Okali, C. 1983. *Cocoa and Kinship in Ghana: The Matrilineal Akan of Ghana.* London: Kegan Paul for the International African Institute.

Okali, C., and J. E. Sumberg. 1985. Sheep and Goats, Men and Women: Household Relations and Small Ruminant Development in Southwest Nigeria. *Agricultural Systems*, 18:39–59.

Okali, C., and J. E. Sumberg. 1986. Examining Divergent Strategies in Farming Systems Research. *Agricultural Administration*, 22:233–253.

Okali, C., J. Sumberg, and J. Farrington. 1994. *Farmer Participatory Research: Rhetoric and Reality.* London: IT Publications.

Okali, C., J. Sumberg, and K. C. Reddy. 1994. Unpacking a Technical Package: Flexible Messages for Dynamic Situations. *Experimental Agriculture*, 30:299–310.

OUP 1992. *Oxford English Dictionary.* Oxford: Oxford University Press (OUP).

Paine, R. 1967. What Is Gossip About? An Alternative Hypothesis. *Man (NS)*, 2(2):278–285.

Paine, R. 1970. Informal Communication and Information-Management. *Canadian Review of Sociology and Anthropology*, 7(3):172–187.

Painter, T., J. Sumberg, and T. Price. 1994. Your "Terroir" and My "Action Space": Implications of Differentiation, Movement, and Diversification for the "Approche Terroir" in Sahalian West Africa. *Africa*, 64(4):447–464.

Pazvakavambwa, S. 1994. Agricultural Extension. In Rukuni and Eicher, eds., *Zimbabwe's Agricultural Revolution.*

Pottier, J. 1994. Agricultural Discourses: Farmer Experimentation and Agricultural Extension. In Scoones and Thompson, eds., *Beyond Farmer First.*

Potts, M. J., G. A. Watson, R. Sinungbasuki, and N. Gunadi. 1992. Farmer Experimentation as a Basis for Cropping Systems Research—A Case Study Involving True Potato Seed. *Experimental Agriculture*, 28(1):19–29.

Prain, G., F. Uribe, and U. Scheidegger. 1992. "The Friendly Potato": Farmer Selection of Potato Varieties for Multiple Uses. In Moock and Rhoades, eds., *Diversity, Farmer Knowledge, and Sustainability*.

Pretty, J. 1991. Farmers' Extension Practice and Technology Adaptation: Agricultural Revolution in Seventeenth–Nineteenth Century Britain. *Agriculture and Human Values*, 8:1–2.

PRATEC (Proyecto Andino de Tecnologías Campesinos). 1991. Andean Agriculture and Peasant Knowledge: Revitalising Andean Knowledge in Peru. In Haverkort, van der Kamp, and Waters-Bayer, eds., *Joining Farmers' Experiments*.

Rappaport, R. 1967. *Pigs for the Ancestors: Ritual in the Ecology of a New Guinea People*. New Haven: Yale University Press.

Ravnborg, H. M. 1992. The CGIAR in Transition: Implications for the Poor, Sustainability, and the National Research Systems. Agricultural Administration (Research and Extension) Network Paper 31. London: ODI.

Rhoades, R. 1989. The Role of Farmers in the Creation of Agricultural Technology. In Chambers, Pacey, and Thrupp, eds., *Farmer First*.

Rhoades, R., and A. Bebbington. 1995. Farmers Who Experiment: An Untapped Resource for Agricultural Research and Development. In Warren, Slikkerveer, and Brokensha, eds., *The Cultural Dimension of Development*.

Rhoades, R. E., and R. H. Booth. 1982. Farmer-Back-to-Farmer: A Model for Generating Acceptable Agricultural Technology. *Agricultural Administration*, 11:127–137.

Rice, R. E., and E. M. Rogers. 1980. Re-invention in the Innovation Process. *Knowledge*, 1:499–514.

Richards, P. 1985. *Indigenous Agricultural Revolution*. London: Hutchinson.

Richards, P. 1986. *Coping with Hunger: Hazard and Experiment in an African Rice Farming System*. London: Allen & Unwin.

Richards, P. 1987. Experimenting Farmers and Agricultural Research. Unpublished manuscript.

Richards, P. 1989. Agriculture as a Performance. In Chambers, Pacey, and Thrupp, eds., *Farmer First*.

Riches, C. R., L. J. Shaxson, J. W. M. Logan, and D. C. Munthali. 1993. Insect and Parasitic Weed Problems in Southern Malawi and the Use of Farmer Knowledge in the Design of Control Measures. Agricultural Administration (Research and Extension) Network Paper 42. London: ODI.

Rijal, D., A. Fitzgibbon, and R. Smith. 1994. How Farmers' Experimentation Can Be Understood. Report for Farming Systems Analysis course, School of Development Studies, University of East Anglia, Norwich.

Robertson, A. F. 1978. *Community of Strangers: A Journal of Discovery in Africa*. London: Scolar Press.

Rogers, A. 1993. Third Generation Extension: Towards an Alternative Model. *Rural Extension Bulletin*, 3:14–16.

Rogers, E. M. 1960. *Social Change in Rural Society*. New York: Appleton-Century-Crofts, Inc.

Rogers, E. M. 1983. *The Diffusion of Innovations*. New York: Free Press.

Rogers, E. M., and F. F. Shoemaker. 1971. *Communication of Innovations: A Cross Cultural Approach*. New York: Free Press.

Röling, N. 1988. *Extension Science: Information Systems in Agricultural Development*. Cambridge: Cambridge University Press.

Roth, M. 1994. A Critique of Zimbabwe's 1992 Land Act. In Rukuni and Eicher, eds., *Zimbabwe's Agricultural Revolution*.

Ruddell, E. 1995. Growing Food for Thought. *Grassroots Development*, 19(1): 18–26.

Rukuni, M. 1994. The Evolution of Agricultural Policy: 1890–1990. In Rukuni and Eicher, eds., *Zimbabwe's Agricultural Revolution*.

Rukuni, M., and C. K. Eicher, eds. 1994. *Zimbabwe's Agricultural Revolution*. Harare: University of Zimbabwe Publications.

Saito, K. A., H. Mekonnen, and D. Spurling. 1994. Raising Agricultural Productivity for Women Farmers and Improving Natural Resource Management: The Cases of Burkina Faso, Zambia, Kenya, and Nigeria. World Bank Discussion Paper 230. Washington, D.C.: World Bank.

Saito, K. A., and D. Spurling. 1992. Developing Agricultural Extension for Women Farmers in Africa. World Bank Discussion Paper 103. Washington, D.C.: World Bank.

Salisbury, R. 1962. *From Stone to Steel*. New York: Melbourne University Press.

Sandford, R. H. D. 1990. *Proposals for a Farmers' Research Project*. Addis Ababa, Ethiopia: Farm Africa.

Schön, D. A. 1983. *The Reflective Practitioner: How Professionals Think in Action*. New York: Basic Books.

Scoones, I., and J. Thompson. 1994a. Knowledge, Power, and Agriculture—Towards a Theoretical Understanding. In Scoones and Thompson, eds. *Beyond Farmer First*.

Scoones, I., and J. Thompson, eds. 1994b. *Beyond Farmer First*. London: IT Publications.

Sharland, R. 1991. A Trap, a Fish Poison, and Culturally Significant Pest Control. In Haverkort, van der Kamp, and Waters-Bayer, eds., *Joining Farmers' Experiments*.

Sims, H., and D. Leonard. 1990. The Political Economy of the Development and Transfer of Agricultural Technologies. In Kaimowitz, ed., *Making the Link*.

Sowers, F., and O. Kabo. 1987. FSR and the Dangling E: Extension, Information Flow and the Farming Systems Approach in Niger. Paper presented at the Farming Systems Symposium, Kansas State University.

Sperling, L. 1992. Farmer Participation and the Development of Bean Varieties in Rwanda. In Moock and Rhoades, eds., *Diversity, Farmer Knowledge and Sustainability*.

Sperling, L., and M. E. Loevinsohn. 1993. The Dynamics of Adoption: Distribution and Mortality of Bean Varieties Among Small Farmers in Rwanda. *Agricultural Systems*, 41(4):441–453.

Sperling, L., M. E. Loevinsohn, and B. Ntabomvura. 1993. Rethinking the Farmer's Role in Plant Breeding: Local Bean Experts and On-Station Selection in Rwanda. *Experimental Agriculture*, 29(4):509–519.

Spring, A. 1988. Using Male Research and Extension Personnel to Target Women Farmers. In S. V. Poats, M. Schmink, and A. Spring, eds., *Gender Issues in Farming Systems Research and Extension*. Boulder: Westview Press.

Stack, J. L. 1994. The Distributional Consequences of the Smallholder Maize Revolution. In Rukuni and Eicher, eds., *Zimbabwe's Agricultural Revolution*.

Stack, J., and C. Chopak. 1990. Household Income Patterns in Zimbabwe's Communal Areas: Empirical Evidence from Five Survey Areas. First Annual Consultative Workshop on Integrating Food, Nutrition, and Agricultural Policy. Harare: University of Zimbabwe.

Stamp, P. 1989. *Technology, Gender, and Power in Africa*. Ottawa, Canada: IDRC.

Stolzenbach, A. 1994. Learning By Improvisation: Farmers' experimentation in Mali. I. Scoones and J. Thompson, eds., *Beyond Farmer First*. London: IT Publications.

Stone, M. P., G. D. Stone, and R. M. Netting. 1995. The Sexual Division of Labour in Kofyar Agriculture. *American Ethnologist*, 22(5):165–186.

Sumberg, J. 1991. NGOs and Agriculture at the Margin: Research, Participation, and Sustainability in West Africa. Agricultural Administration (Research and Extension) Network Paper 27. London: ODI.

Sumberg, J., and M. Burke. 1991. People, Trees, and Projects: A Review of CARE's Activities in West Africa. *Agroforestry Systems*, 15:65–78.

Sumberg, J., and C. Okali. 1988. Farmers, On-Farm Research, and the Development of New Technology. *Experimental Agriculture*, 24(3):333–342.

Swift, J. 1979. Notes on Traditional Knowledge, Modern Knowledge, and Development. *IDS Bulletin*, 10(2):41–43.

Trigo, E. J. 1987. Agricultural Research Organisation in the Developing World: Diversity and Evolution. In V. W. Ruttan and C. E. Pray, eds., *Policy for Agricultural Research*. Boulder: Westview Press.

van Beek, W. E. A. 1993. Processes and Limitations of Dogon Agricultural Knowledge. In Hobart, ed., *An Anthropological Critique of Development*.

Van der Ploeg, J. D. 1989. Knowledge Systems, Metaphor and Interface: The Case of Potatoes in the Peruvian Highlands. In Long, N., ed., *Encounters at the Interface: A Perspective on Social Discontinuities in Rural Development*. Wageningse Socioloisches Studies 27, Wageningen: University of Wageningen.

Van der Ploeg, J. D. 1990. *Labour Markets and Agricultural Production*. Boulder: Westview Press.

Van der Ploeg, J. D. 1993. Potatoes and Knowledge. In Hobart, ed., *An Anthropological Critique of Development*.

Vel, J., L. V. Veldhuizen, and B. Petch. 1991. Beyond the PDT Approach. In Haverkort, van der Kamp, and Waters-Bayer, eds., *Joining Farmers' Experiments*.

Versteeg, M. N., and V. Koudokpon. 1993. Participative Farmer Testing of Four Low External Input Technologies to Address Soil Fertility Decline in Mono Province (Benin). *Agricultural Systems*, 42(3):265–276.

Waibel, H., and D. Benden. 1990. Farmers as Members of Research Teams in Experiment Stations. In L. Cammann, ed., *Peasant Household Systems*. Berlin: Deutsche für Internationale Entwicklung (DSE).

Waldie, K. J. 1995. Wenchi Farmer Innovation Survey, Awisa Report. Unpublished report of the Wenchi Farming Systems and Training Project, Wenchi, Ghana.

Wallace, I., K. Mantzou, and P. Taylor. 1996. Policy Options for Agricultural Education and Training in Sub-Saharan Africa: Report of a Preliminary Study and Literature Review. Working Paper 96/1. Reading: Department of Agricultural Extension and Rural Development, University of Reading.

Waters-Bayer, A. 1988. Soybean Daddawa: An Innovation by Nigerian Women. *ILEIA Newsletter*, 4(3):8–9.

Wellard, K., and J. K. Copestake, eds. 1993. *State-NGO Interaction in the Development of New Agricultural Technology for Small Farmers: Experiences from Sub-Saharan Africa*. London: Routledge.

Wilson, K. B. 1990. *Ecological Dynamics and Human Welfare: A Case Study of Population, Health, and Nutrition in Southern Zimbabwe*. Ph.D. diss., University College London, London.

Winarto, Y. 1994. Encouraging Knowledge Exchange: Integrated Pest Management in Indonesia. In Scoones and Thompson, eds., *Beyond Farmer First*.

World Bank. 1992. *A Strategy to Develop Agriculture in Sub-Saharan Africa and a Focus for the World Bank.* Washington, D.C.: World Bank.

World Bank. 1994. Agricultural Extension: Lessons from Completed Projects. Report 13000, Operations Evaluation Department. Washington, D.C.: World Bank.

Index

Africa: agricultural population distribution by agro-ecological zone, 22, 23T; development and poverty alleviation policies in marginal areas, 21–24; export-commodity research focus in, 12, 16; maize research in, 16; national research systems in, 10–11, 12–14; nongovernmental organizations in, 23–24. *See also* Farmers' experiments, Africa

African agricultural extension systems, 16–21, 23: constraints on, 20–21; criticisms of, 18, 21, 43, 125, 159–160; improvements in, 67–68; in marginal areas, 23; T&V (Training and Visit) systems in, 17, 51, 67

Agricultural extension systems: arguments for greater openness and flexibility in, 160–161; in developing countries, criticisms of, 159–160; and diffusion of innovations research, 33; in East Anglia, 85; and facilitation of experimentation, 43; information exchange and knowledge creation in, 17, 18–19, 51; as inhibiting factor in experimentation, 153; in marginal areas, 22–23; NGOs' expanding role in, 20–21; participative feedback models, 19–20; and research linkages, 16, 19, 43; second generation model, 19; stereotypical patterns of organization behavior in, 18–19

Agricultural research, formal: and agricultural extension system, 16, 19, 43; argument for greater farmer participation in, 4–6; client-orientated improvement initiatives in, 13–14; commodity-oriented, 43, 153; common knowledge analysis of, 11–12, 15; criticism of, 10–12; effectiveness and impact of, 5, 10, 15–16; and farmers' experimentation, synergy versus complementarity in, 154–155, 157; and farmers' experimentation linkages, 38–39, 48–49, 151; and farming systems research, 13–14, 19; history of farmer participation in, 14–15; as incremental

improvements to established practices, 107–108; and "indigenous" research, 98, 108–109; ineffectiveness of normative models, 5; positive return to investment in, 10; poverty and development agenda in, 21–24, 159; practical versus theoretical orientation of, 37; in research literature, 9; technology transfer subsystem of, 10, 16. *See also* African agricultural extension systems

Agricultural technology systems, defined, 10

CGIAR. *See* Consultative Group for International Agricultural Research

Cocoa Research Institute of Ghana (CRIG), 56

Consultative Group for International Agricultural Research (CGIAR), 10–11, 12–13; regional "centers of excellence," 12–13

CRIG. *See* Cocoa Research Institute of Ghana

Diffusion of innovations theory: and agricultural extension, influence on, 33; and farmers' experiments, 41–42, 119–120; and "reinvention," 120; and transfer of technology approach, 33

East Anglia, United Kingdom. *See* Farmers' experiments, East Anglia

Environmental degradation, 43–44, 153

Experimentation, defined, 1, 147

Export commodities, institutional focus on, 12, 16

Farmer participatory research: in current development discourse, 3; and degree of intervention level, 27; diverse approaches and activities in, 24–27; empowerment objective of, 6, 27, 28, 154; as model of agricultural extension, 158; proactive versus reactive experiments in, 41; as related to farmers' experimenta-

tion, 34; research-driven versus development-driven, 26–27, 157–160; social development agenda in, 6, 25, *26*; synergy hypothesis in, 27–30, 154–155; technical concerns agenda in, 25, *26*
Farmers: empowerment of, and participatory experimentation, 6, 27, 28, 154; as experimenters, categories of, 45–48; as innovators versus early adopters, 46; research-minded, 119–121, 150, 154, 157, 159
Farmer's experiments: as adaptive research, 152; characteristics of, 37–40; as embedded in local environment, 39, 40–41; and formal research linkages, 38–39, 151; and increasing hegemony of formal research, 108–109; as incremental improvements to established practices, 107–108; as informal research system, 48–49; level of, and degree of diversification, 153; level of, and environmental degradation, 43–44, 153; limited benefits of, 155–156; and local traditions, 151–152; outputs and processes of, 35–37, 38; as planned and purposeful, 98, 108; predominance of technical experiments in, 151; prevalence of, 36–37; and private versus public domain, 39–40; site-related factors in, 150; sociocultural and socioeconomic factors in, 44–45, 150, 153; typologies of, 40–41; as untapped development source, 155–156
Farmers' experiments, Africa: characteristics of, 87–103; classification of experimentation in, 58, 59T; and commitment to farming, 122–123, 154; and competition, 126; controls and comparisons in, 98–100; and crop propagation method, 124; effects of adverse conditions on, 92–93, 121; and farmers' characteristics, 60–62; and farmers' perceptions of innovators, 126–133, 144–145; and farmers' willingness to discuss innovation, 126–142; increased commercialization as stimuli to, 126; influence of formal extension activities on, 101; and life cycle of crops, 124–125; and local information networks, 126–142; and openness of landscape, 123–124; proactive character of, 96–98; profiles of experimenters in, 111–119; and promotional activities of extension programs, 125; research methods, 56–63; research sites, 55–56; as response to shared perceptions of prob-

lems, 121; sources of ideas and techniques in, 100–103; topics of experimentation in, 89–96; trialability and risk in, 95–96, 126. *See also* Ghana; Kenya; Zimbabwe
Farmers' experiments, East Anglia: adaptive and applied research in, 148; and agricultural extension services, 85; characterization of outputs in, 107; formal research systems, 85; global and environmental impacts on, 84–85; and Ideas Competition in farm machinery, 62–63, 103, 142–143; and major crops of area, 84; and Morley Research Centre, 85; origin and ownership of, 107; as proactive and reactive, 105; and scale of farming enterprise, 143; side-by-side comparisons in, 105–107; topics of, 103–104
Farming systems: anthropological and ethnographical studies of, 32–33; European and North American development of, 32
Farming systems research, 13–14, 19–20

Gender roles, 44–45, 116–119, 150
Ghana: Brong Ahafo Region fieldwork site, 60, 61, 80–84; cocoa production in, 76; development programs in, 79; Eastern Region fieldwork site, 57, 61, 76–79; Economic Recovery Programme in, 67; food crops, 67, 80, 81–82, 83; nature of agricultural sector in, 66; nonfarm income in, 76–77; research and extension systems in, 67, 83–84; socioeconomic characteristics of experimenters in, 111, 112, 116; tomato crop experimentation in, 79, 81–82, 83, 93–94, 124, 126, 130–131, 138–139; traditional crops, 77–78
Global 2000 program, 82
Green Revolution, 4, 10, 21

ICIPE. *See* International Centre for Insect Physiology and Ecology
IITA. *See* International Institute for Tropical Agriculture
ILCA. *See* International Livestock Centre for Africa
Innovators: categories of, 130–133; defined, 119; versus early adopters, 46; and research-minded farmers, 119–121
Intermediate Technology Development Group (ITDG), 55–56, 57, 76, 93, 101, 125, 129

International Centre for Insect Physiology and Ecology (ICIPE), 55
International Institute for Tropical Agriculture (IITA), 3
International Livestock Centre for Africa (ILCA), 3
International research centers, 10–11, 12–13
International Service for National Agricultural Research (ISNAR), 11, 13
ISNAR. *See* International Service for National Agricultural Research
ITDG. *See* Intermediate Technology Development Group

Kenya: impact of colonial, white settler economy in, 65–66; Kangare village fieldwork site, 56, 60, 61, 71–73; national research and extension systems in, 67; structural adjustment policies in, 66

Local communication networks, 49–52, 126–142; extension services' use of, 51; farmers' willingness to discuss innovation in, 126–133; free flow of information in, 137–139; international sources of information in, 145; role of formal and informal groups in, 139–142; trust and gossip in, 51–52
Local environment, embeddedness of farmers' experiments in, 39, 40–41
Local knowledge: defined, 1; farmer-to-farmer transfer of, 32; versus formal knowledge in, 98, 108–109, 149–150; and second generation extension models, 19

Migration, 44

NGOs. *See* Nongovernmental organizations
Nongovernmental organizations (NGOs), 3; extension activities of, 20–21; poverty and social development agenda of, 14, 23–24
Norfolk Farm Machinery Club (NORMAC), Ideas Competition, 62–63
NORMAC. *See* Norfolk Farm Machinery Club

Poverty: and farmers' experimentation, 42, 43, 119, 153; as new agenda of agricultural research, 21–24, 159

Reinvention, farmer's experimentation as, 45
Research-minded farmers, 119–121, 150, 154, 157, 159
Resettlement, 44
Risks of experimentation, 42–43, 95–96, 126
Rwanda, national bean breeding program in, 157

Social development agenda, 14, 23–24, 25, 27, 28
Structural adjustment programs, 12, 66, 75–76
Synergy hypothesis of participatory research, 27–30, 154–155

T & V systems. *See* Training and visit systems
Technical experiments, predominance of, 151
Technology transfer, as agricultural research subsystem, 10, 16
Tomato crop experimentation (Ghana), 79, 81–82, 83, 93–94, 124, 126
Training and Visit (T&V) systems, 17, 51, 67

Wenchi Farm Institute (WFI), 56
Wenchi Farming Systems and Training Project (WFSTP), 56, 82
WFI. *See* Wenchi Farm Institute
WFSTP. *See* Wenchi Farming Systems and Training Project
World Bank, 11, 13, 21, 66; Training and Visit (T&V) systems funding, 17
World Vision, 79

Zimbabwe: agricultural extension system in, 68; Chivi District fieldwork site, 56–57, 60, 61, 73–76; development and structural adjustment policies in, 66, 75–76; effects of colonial settler economy on, 65; experimenters' socioeconomic characteristics in, 112, 115–116; Farmers Clubs and Master Farmer Training Scheme in, 68, 75, 125, 136, 137–138, 142; second agricultural revolution in, 66–67

About the Book

Since the mid-1970s, growing interest in greater farmer participation in formal agricultural research has had major implications both for investment priorities and for models of organization, implementation, and management of agricultural research and development.

Sumberg and Okali identify, characterize, and contextualize the experimental activities undertaken by farmers themselves, providing a theoretical and empirical base from which alternative models for the interaction of formal research and farmers' experiments can be evaluated. Their work is a seminal contribution to the debate among donors and development agencies, as well as in academic circles, about the search for more socially and economically acceptable models for the development and transfer of agricultural technology.

JAMES SUMBERG is senior lecturer and CHRISTINE OKALI is research fellow in the School of Development Studies at the University of East Anglia, United Kingdom.